FUTURE FIRE

"It is time for us to start a buildup. And it is time for us to build up to a point where no other nation on this earth will ever dare raise a hand against us. In this way we will preserve world peace."

President Ronald Reagan

FUTURE FIRE

Weapons for the Apocalypse

A PRINT PROJECT BOOK

by Ann Marie Cunningham
& Mariana Fitzpatrick

WARNER BOOKS
A Warner Communications Company

Warner Books, Inc., 666 Fifth Avenue, New York, N.Y. 10103

A Warner Communications Company

Printed in the United States of America

First printing: June 1983
10 9 8 7 6 5 4 3 2 1

Book design: Judy Allan (The Designing Woman)

Library of Congress Cataloging in Publication Data
Main entry under title:
Future fire.

"A Print Project book."
Includes index.
1. Weapons systems. 2. Arms and armor. 3. War.
4. Military art and science. I. Cunningham, Ann Marie.
II. Fitzpatrick, Mariana.
UF500.F87 1983 355.8'2 82-17381
ISBN 0-446-37031-2 (USA)
ISBN 0-446-37314-1 (Canada)

This book is dedicated to the spirit of journalistic inquiry, and to the many journalists whose work we've reported herein. That they are uncovering the facts underlying Future Fire *is crucial to life in the 1980s and 1990s, and perhaps to survival itself.*

The Print Project

contents

> ". . . unless another war is prevented it is likely to bring destruction on a scale never before held possible and even now hardly conceived, and little civilization would survive it. . . . Our situation is not comparable to anything in the past. We must revolutionize our thinking, revolutionize our actions, and must have the courage to revolutionize relations among the nations of the world. . . . We must overcome the horrible obstacles of national frontiers."
>
> Albert Einstein
> *Out of My Later Years*

introduction

Weapons are centrally related to defense policy. What we can do, or what we think we can do, with our neutron bombs and binary nerve gases, our "killer" satellites and space battle stations, in large degree determines the stance we take in relation to the rest of the world. Thus, for example, the increasingly refined targeting capabilities of directed energy weapons like lasers and particle beams informed President Carter's conviction that we will soon be capable of waging "a limited but protracted" nuclear war. No longer reliant upon those crude megabombs that flatten everything within miles of Ground Zero, we can begin to selectively zap Russia's munitions plants, her schools, her centers of government. Nuclear pinpointing capability! The concept suggests that we can enter into a "limited" (and thus, presumably, predictable) nuclear exchange, going carefully back and forth, with us "pinpointing" something of theirs, and them "pinpointing" something of ours.

Calling Presidential Directive 59 "apocalyptic nonsense," Paul N. Warnke, President Carter's former arms control and disarmament director, has warned that countries equipping themselves "to fight a limited nuclear war will end up fighting an all-out nuclear war that nobody could win." In war, things always escalate beyond the original plan. The tit-for-tat development of new weapons by the superpowers leads to less security and greater hair-trigger volatility—not only among the superpowers but among frightened developing nations who are desperate to get as much nuclear capability as they can, as quickly as they can.

Technologically, it's a new kind of war we're talking about. Both the United States and the USSR are close to being able to orchestrate whole battles simply by pushing buttons. Computers have revolutionized defense. They design. They test. They plan strategy and make decisions. They warn (or fail to). They "read," "see," and "hear."

One important aspect of advanced weapons technology is speed. The refinement of nuclear weaponry has been steadily decreasing the interval in which alternative defense options can be considered. For example, the replacement of liquid fuel by solid fuel means rockets can stand in their silos, instantly ready. The time of delivery has contracted. In the mid-1970s the time required for the interhemispheric delivery of nuclear bombs had shrunk to about ten minutes. Now it is much less.

Another important factor in the new defense thinking is the directed energy weapon, made either of charged or neutral particles (the new "lightning bolt" weapon), or of photons of light (the new laser weapons). The range of these new beam weapons, when fully developed, will be extraordinary: thousands of miles, traveling at close to the speed of light (about 186,000 miles a second).

Beam weapons have led to the exquisite if dubious concept that the dirty work of international dispute might be accomplished in space, with nary a human being singed. Should Russia push the button to send up her missiles, our satellite-based laser and lightning-bolt weapons, with their sophisticated tracking and sensing equipment, would spot those missiles and instantly destroy them. A smashing display of thermonuclear pyrotechnics would occur thousands of miles out in space—only, as with the tree falling in the forest, no one would be there to see it. The effects of that attempt at a preemptive first strike—and the aborting antistrike—would be witnessed in small patterns on cathode-ray tubes in generals' offices down below. As in the computerized war games kids play in penny arcades, those CRTs would blip messages of space disaster in small white dots and rectangles, images of silent toy missiles, and zigzag bolts of artificial lightning. Space war, the Department of Defense (DOD) posits, could be clean and pure. Increased "lethality" and decreased human carnage have become concepts in the defense credo for the future.

To spend time contemplating so pristine a vision of war in space is a dangerous distraction. We already have the technology for devastating attacks on or near the surface of the earth, or below the level of the sea, on submarines longer than the Washington Monument is tall. Today's nuclear bombs are a thousand times more powerful than the bomb dropped on Hiroshima. All told, the nuclear explosive potential in the world today is a million times the power released by that first bomb. Plainly we are working with exponentially greater levels of harnessed energy than ever were contemplated by the Los Alamos scientists who invented the A-bomb.

Everyone with a vote to cast and a brain to evaluate needs to know how the power of the new MEGAWEAPONS relates to our lives. This is a primer of modern weapons technology. It describes the latest "hardware" (what the apocalyptic new weapons are, how they work, what they're capable of), and it discusses the political issues created—and to some extent determined—by the technology. Unrestrained nuclear proliferation, the long-term effects on the planet of radioactive waste, the increasing possibility of grotesque nuclear accident (including the ultimate accident, known to the weapons cognoscenti as NUCFLASH)—these are the questions ordinary citizens can no longer ignore. Make no mistake. It has been a policy of the government—in particular the departments of defense and energy—to keep us in the dark. In the dark and hoping. Hoping the politicians will make the right decisions. Hoping the government will exercise its power responsibly. Hoping the nuclear nightmare—that awful dream of the fireball, which even now is invading the sleep of our children—will not return in reality, turning life itself to ash.

Today's decisions about weapons impose the political choices of tomorrow. Today's decisions about weapons determine whether, in fact, there will be a tomorrow.

The Print Project has put together this book in the hope that piercing the mystique of MEGAWEAPONRY will help people come to grips with our increasing responsibility for controlling technology.

"Now there are some fifty thousand warheads in the world, possessing the explosive yield of roughly twenty billion tons of TNT." —Jonathan Schell, The Fate of the Earth.

chapter 1

The Road to Megaweapons: *The History of U.S. Nuclear Strategy*

Americans seem out of touch with the fact that the power of the bombs dropped on Hiroshima and Nagasaki in 1945 were trifling compared to the devastating power of the weapons of the 1980s. Any modern hydrogen bomb would yield at least a thousand times the deadly blast, heat, and radiation unleashed on Japan. "If we and the Soviets were to let everything fly at once," Lewis Thomas has written, "we could do, in a matter of minutes, a million times more damage than was done on those two August mornings long ago." Twelve miles in every direction from the detonation, nothing would survive, no matter how bomb-sheltered or "hardened" (covered with massive concrete shield to withstand the blast).

The Soviet Union and the United States are not the only nations with such terrible power. Four other countries—Great Britain, France, China, and India—also have nuclear weapons. Nine more could build them now if they chose to. Two, Israel and South Africa, may have already done so. Sixteen other nations, including such volatile countries as Libya, Iraq, Egypt, South Korea, and Argentina, could build a nuke within the next ten years.

Some of them will undoubtedly do so, using nuclear technology imported for the "peaceful" purpose of building reactors.

The Israeli bombing of the Iraqi reactor that may have been producing bomb materials demonstrated just how weak are the international controls on nuclear bomb-making materials. The thirty-five-year-old international effort to curb the spread of nuclear weapons has been drastically undermined by the export—at first by the United States alone, now other countries as well—of domestic nuclear technology. The Israeli raid and President Reagan's plan to extract bomb material from recycled waste from nuclear reactors make clear that domestic and war-making nuclear technology are ultimately one and the same. If a country builds a reactor, it can one day build a bomb. And the larger the membership in the nuclear club, the greater the risk that one day the deadly weapons will be used. **Some intelligence sources estimate that the critical nuclear confrontation will arrive as early as 1985.**

A worse threat than outright nuclear war is the possibility of a terrorist group acquiring nuclear materials, building a football-

The bombing of Hiroshima. What happened here represents one millionth the total explosive power now available in the world.

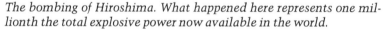

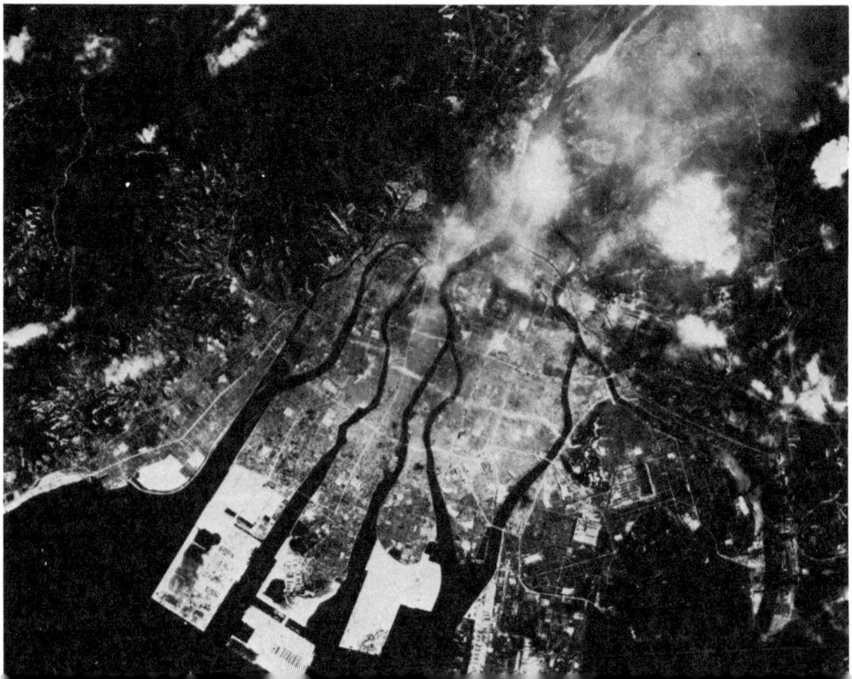

size Hiroshima bomb, and holding a major city or even a country hostage. Bombs are relatively simple to build. All that impedes terrorists is the fact that it takes a long time to collect the necessary plutonium or enriched uranium from conventional reactors. But Reagan is encouraging the building and export of breeder reactors, which produce more bomb material faster.

The arms race is going full tilt. New weapons are constantly being perfected as the search for the ultimate deterrent continues—and there are huge numbers of nukes. At the current rate of manufacture we can expect to find ten thousand new atomic weapons in place around the world within the next ten years. The one sure method of limiting the arms race—severely curtailing research on and development of new weapons—is stymied by the passion on both sides for technological advance.

When scientists spot something "technically sweet," they tend to go ahead and develop it and think about the consequences afterward, said J. Robert Oppenheimer, the physicist who headed the Manhattan Project's Los Alamos lab, which developed the atomic bomb in 1945. The scientists who invented nuclear weapons were the first to become concerned about stockpiling them. After Hiroshima and Nagasaki these scientists, including Oppenheimer, worked hard to alert politicians and the public to the dangers of the arms race. In the late 1950s and early 1960s, before bomb tests were banned aboveground, the public shared the scientists' concern. But as bomb testing went underground, so did people's nightmares. For twenty years we learned to live with the bomb, even if we couldn't love it. But suddenly during the presi-

The first atomic bombs were produced at Los Alamos Scientific Laboratory, in New Mexico. After witnessing the first test explosion, bomb design chief Robert Oppenheimer said, "We felt the world would never be the same again."

dential campaign of 1980 it became clear that both candidates considered the idea of "limited nuclear exchange" an acceptable possibility. Evidently there has been a major shift in strategic thinking. Some politicians now talk about nuclear war as they would about any other kind of war they could win or lose.

THE ROAD TO LOS ALAMOS

How did we come to such a pass?

The road to today's megaweapons started at Los Alamos, New Mexico, the site of the Manhattan Project's main laboratory, where atomic weapons were first invented and fabricated. Why did those scientists sign up for such a deadly project in the first place?

Ever since radioactivity was discovered, scientists have envisioned atomic utopia—as well as incredibly destructive weapons. Frederick Soddy, codiscoverer of radioactivity, predicted in 1904 that since modern civilization rests on a ready supply of energy, a nation that harnessed the power of the atom "would have little need to earn its bread by the sweat of its brow. Such a race could transform a desert continent, thaw the frozen poles, and make the whole world one smiling Garden of Eden." Soddy also predicted that anyone who learned to control nuclear energy "would possess a weapon by which he could destroy the earth."

In 1913 the ever-prescient H. G. Wells wrote a science fiction novel (it became a 1936 film, *Things to Come*) in which he coined the phrase "atomic bombs." He visualized a 1950s world war in which these bombs wiped out civilization. But afterward a world government would be set up to keep the miraculous new energy free of petty national interests. As Wells's novel ends, the world is prosperous and peaceful, due to abundant nuclear energy.

By the 1930s scientists all over the world were hard at work harnessing the power of the atom. They were well aware of its potential destructiveness but were much more excited by the thought of making Soddy's Garden of Eden a reality. Germany, especially, was making brilliant scientific advances. But when Adolf Hitler took power, in 1933, he forced Jewish scientists,

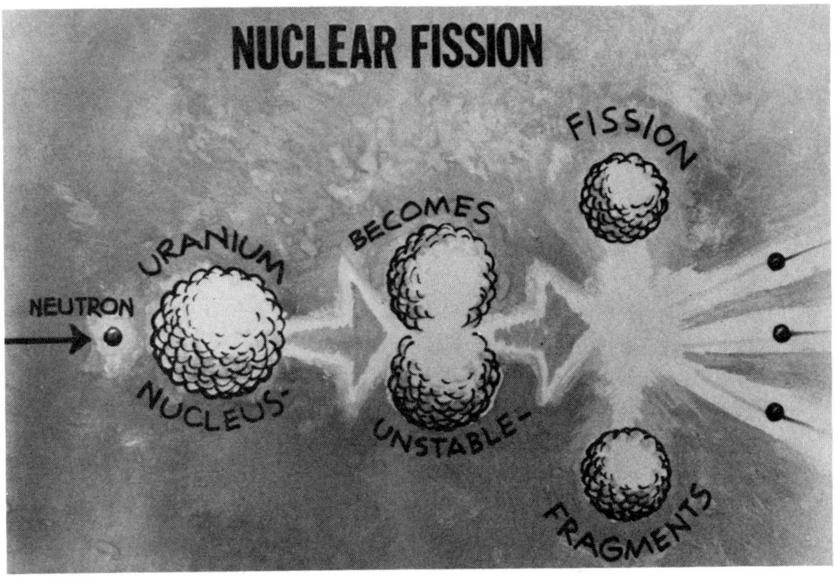

A historic experiment in 1938 showed that the nucleus of the heaviest known atom—uranium—would split in two when struck by a neutron. This process, known as fission, *releases energy that's 100 million times greater—per atom—than the energy of burning oil.*

some of whom were the greatest in modern physics, out of university jobs. By the outbreak of World War II most of Germany's Jewish physicists, including Albert Einstein himself, had come to the United States, while their non-Jewish colleagues stayed behind. Emigré scientists, worried about both the spread of Nazism and the prowess of German science, persuaded a willing Einstein to write President Franklin D. Roosevelt telling him it was crucial that America beat Germany to the bomb. As a result of Einstein's communiqué, the Manhattan Project was born. Under Oppenheimer's direction the Los Alamos Scientific Laboratory, birthplace of the bomb, opened in 1942. It is still going strong.

Initially, fear of the Nazis was the prime mover behind the U.S. nuclear weapons program. As Oppenheimer said later, "Almost everyone realized that this was a great undertaking." The scientists he recruited, among the most intelligent and idealistic

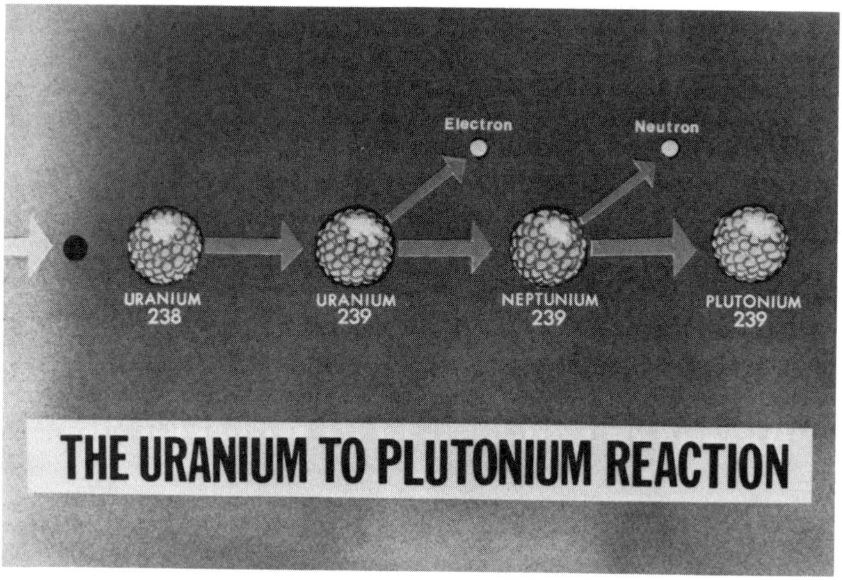

An atom of plutonium is made when a nucleus of Uranium-238 captures a neutron. This happens continually during the operation of nuclear reactors.

members of their generation, believed that one way or another, harnessed nuclear energy would mean a better future. The bomb would be so powerful it would protect forever the country that owned it; war would be outlawed.

This was an appealing project to the young scientists at Los Alamos, most of whom were graduate and postgraduate students. There's also no doubt that building the bomb was a unique career opportunity, a chance to work with the best minds from all over the world, with the best equipment, supported by massive government funding. Some physicists (Freeman Dyson is one) believe that at Los Alamos science made a Faustian bargain with the federal government, agreeing to develop the bomb in exchange for enormous power and professional rewards. That may be overly simplistic (scientists can rarely foresee every social consequence of their research and development), but there's no doubt that at Los Alamos, Big Government made Big Science. It's significant that after VE Day bomb research at Los Alamos didn't

Here, held in gloved hands, is a plutonium "button"—enough to make a nuclear bomb. Manufactured in special plants (one is in Hanford, Washington, another in Savannah River, Georgia), plutonium is deadly stuff. Even microscopic amounts can cause lung cancer when inhaled as gas or dust.

slow down at all. Eventually the Manhattan Project was to become the grandparent of the space shots, the fight against cancer, the development of peaceful nuclear energy.

Once Hitler was defeated, some Manhattan Project scientists began having qualms, even before the first test of the bomb in New Mexico. They objected to the A-bomb not only on moral grounds but because they realized that it would result in a huge postwar arms race. In the end the scientists didn't have sufficient influence to prevent the government from using the bomb on Japan. President Harry S Truman may have worried about how he could otherwise justify to Congress the staggering $2 billion price tag on the bomb. In the end we may have used the first atomic bomb simply because it was there.

In *The Nuclear Barons,* Peter Pringle and James Spigelman argue that the scientists didn't use their real trump card—the long-term hazards of radiation—because they were naive enough to assume it had no political relevance. They knew that many Japanese would die from the immediate radiation from the bomb's blast, and that many more would die in years to come from delayed effects. But the scientists didn't tell Truman about the delayed effects of radioactivity.

The bombings of Hiroshima and Nagasaki were actually less *immediately* disastrous than the allied fire bombings of Dresden and Tokyo, which destroyed more property and took more lives. But the tragedy of Hiroshima and Nagasaki continues because at least one hundred Japanese die every year from the aftereffects. If Truman had been told the facts about radiation, he might have understood that dropping that first bomb had devastatingly far-reaching consequences.

Artist's rendering of the primitive reactor in which the first fission chain reaction was produced.

As it was, both the military and the public cheered after Hiroshima and Nagasaki. The war was over. Hiroshima and Nagasaki introduced the nuclear age, but their ruination also crowned America as *the* global power. After the war some atomic scientists continued to try to warn the public about the dangers of the bomb. They sent lumps of fused sand from the New Mexico test site to the mayors of 42 cities, just to remind them what could happen. At the same time, the scientists pushed for the development of peaceful nuclear power. As David Lilienthal, the first head of the Atomic Energy Commission (AEC), which was charged with developing domestic nuclear power, recalled, "Somehow or other, the discovery that produced so terrible a weapon simply had to have an important peaceful use."

THE AGE OF ASSURED ASCENDANCY

The question was how to outlaw a weapon while encouraging the spread of its technology for peaceful purposes. Never before had a nation faced this problem, and judging by the current proliferation of nuclear weapons, we haven't solved it yet. But probably we wouldn't even have tried had we not felt comfortably ahead in the arms race.

The first chapter of post–World War II nuclear strategy has been labeled "The Age of Assured Ascendancy." The phrases "limited nuclear war" and "first strike," which caught the public's attention when they surfaced as part of the campaign rhetoric in 1980, are not new. The Pentagon first began to contemplate these concepts in the late 1950s, once it had become worried that the United States was losing its edge in the arms race. Right after World War II, when we thought the Russians wouldn't have their own bomb for another twenty years, military planners considered the bomb the only necessary deterrent. If another world war were fought, the bomb would be used against enemy cities along with conventional weapons. But Henry Kissinger, then an obscure political science professor at Harvard, warned that the United States had added the atomic bomb to its arsenal "without integrating its implications into military thinking."

In 1949 the Soviets exploded their first atomic device. So much for the atomic bomb as the ultimate deterrent. It was a rude shock to the United States to learn that, due to the work of a spy at Los Alamos, the Russians were far ahead of schedule. If we were to keep pace in the race, we would have to start work on the hydrogen bomb.

The Soviet bomb made clear that no nation was going to forgo nuclear weapons just because the United States didn't like the idea of others having them. Building a nuclear bomb meant graduating to modern statehood; it became a matter of national pride. In 1952 Great Britain joined the nuclear club, expanding the membership to three and forcing the United States to recognize that if we ever bombed the enemy again, we'd be bombed right back. At this point military strategists began speaking of "national security" instead of "national defense."

In 1954 President Dwight D. Eisenhower inaugurated the concept of *massive retaliation*, the first official defense strategy to recognize nuclear bombs as something other than ordinary weapons. If the Russians or the Chinese Communists breached the peace anywhere, the United States would respond by threat-

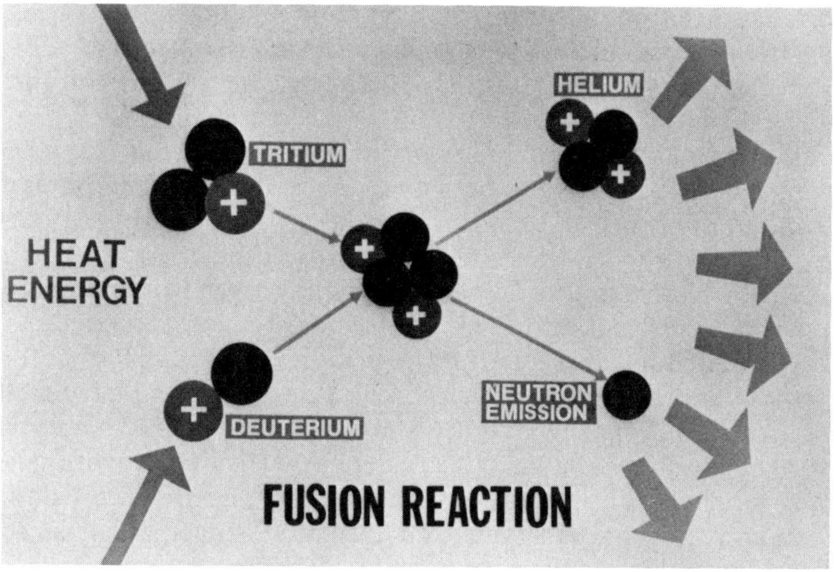

Fusion *is the basic reaction of the hydrogen bomb. Basically it's the opposite of the* fission *reaction, in which atomic nuclei are split. In fusion the nuclei of light elements (tritium and deuterium) are joined, or fused, to create a heavy element (helium). Neutrons of free energy are emitted when the fusion takes place, and it is this free energy that empowers the hydrogen bomb.*

ening to attack the aggressor's homeland with nukes. The U.S. Air Force's Strategic Air Command (SAC) was given responsibility for carrying out massive retaliation. General Curtis LeMay created a nuclear air force: nuke-armed "looking-glass planes" that would always be in the air, long-range intercontinental ballistic missiles (ICBMs), land-based in silos hardened to withstand nuclear blast, and command headquarters in a hollowed-out mountain near Omaha, Nebraska, in constant readiness to effect a strike anywhere.

Massive retaliation was considered part of an earlier policy, *deterrence,* which the Pentagon still officially follows. As such, massive retaliation is presented as a defensive measure because it is a second nuclear strike in response to a first strike by the other side. The capacity for second strike—massive and unacceptable retaliation—theoretically deters the Russians from attacking us in the first place. In order for massive retaliation to be an effective deterrent, however, U.S. forces have to be able to survive the worst conceivable enemy attack, the destruction of every single one of our 1,052 land-based ICBMs, and *still* wreak massive destruction on the Soviet Union. This is why our long-range nuclear forces comprise a triad: SAC planes armed with nukes; strategic missiles in their hardened silos; and strategic missiles on submarines. Two parts of the triad, SAC planes and subs, both relatively hard to trace and therefore immune from attack, would be used to retaliate.

The massive-retaliation strategy, however, had a loophole. Because the United States couldn't responsibly respond to small infractions with nukes, the other side began to conduct its offensive through small bites, each safely below a level that would justify massive retaliation. The world survived the Cuban missile crisis—deterrence seemed to work there—but meanwhile the Russians were building their own intercontinental ballistic missiles.

In the late 1950s Herman Kahn, founder of the Hudson Institute, a hard-nosed think tank, began thinking the unthinkable. He argued that nuclear war should be considered seriously as a last resort, winnable if we were strong, for he believed that the human race could survive. In 1957 Henry Kissinger had come to national attention by publishing *Nuclear Weapons and Foreign Policy,* in which he coined the phrase, "limited war," advocating the use of tactical nukes. The problem with limited nuclear war was that no one could think of any practical way of keeping it limited. There was no guarantee that conflict wouldn't quickly escalate out of control. In 1961 Kissinger published what amounted to a retraction of his ideas on limited war, but no military strategists were paying attention to him. In the early 1960s they refined the concept of massive retaliation with a new, double-pronged strategy. The first prong was the concept of *flexible response,* which called for (1) renewed emphasis on conventional Western military (nonnuclear) strength; (2) declared willingness to "escalate as necessary" to turn back aggression; and (3) the "forward placement" of tactical (short-range) nuclear weapons on the battlefield.

There remained the problem of how to prevent a strategic (long-range) attack on the United States. The answer was the second new prong of massive retaliation: *damage limitation.* It meant the United States could go ahead with a "preemptive first strike" to disable threatening Soviet forces before they had a chance to attack. A preemptive use of strategic weapons to disable an enemy's nuclear force is not considered the same as "first use," which means that a country will initiate the employment of nuclear weapons in war. "First use" was, and remains, part of *flexible response.* Of the current members of the nuclear club, only China has formally pledged never to be the first to use nuclear weapons. Thus, with *flexible response* and *damage limitation,* the United States was abandoning the defensive posture of *massive retaliation* as deterrence and reserving the right to wage limited war with a first strike. Deterrence was officially rechristened *assured destruction* and, as Soviet forces continued to grow to a standoff, *mutual assured destruction,* or MAD.

Throughout the 1960s the United States stayed comfortably on the plus side of the MAD equation. MAD actually seemed to work as a stabilizer. Neither the United States nor the USSR could attack the other without risking a fatal attack on itself. Oppenheimer once compared the two world powers to "two scorpions in a bottle." But as Soviet nuclear forces increased to equal ours, in the early 1970s, the problems of conducting a so-called limited first strike on the enemy's missiles multiplied. The age of Assured Ascendancy had given way to the Age of Assured Anxiety.

Atomic bombs.

THE AGE OF ASSURED ANXIETY

In 1969 U.S. nuclear strategists began to reassess their position. It had become clear that the Soviets were at least as strong as, if not stronger than, the United States. President Richard M. Nixon raised a chilling question: "Should the concept of assured destruction be narrowly defined, and should it be the only measure of our ability to deter the variety of threats we might face?"

Nixon worried about the limited possibilities of MAD because the U.S. response to a nuclear attack was supposed to destroy the enemy's cities and industries rather than its unexpended strategic forces. Though MAD was meant to cripple and demoralize the aggressor, many strategists considered the policy inhumane and downright stupid. Why should the United States strike at people and industry rather than limit a destructive nuclear war by knocking out unused missiles?

Accordingly, MAD was mated with a *flexible strategic targeting* policy, which meant that we could aim at military targets in response to a limited Soviet strategic attack, or to a Soviet attack on U.S. allies that tactical nukes hadn't stopped. This combination is variously known as *assured retaliation, counterretaliation,* or *counterdeterrence.* The policy's new component, the destruction of Soviet missile silos, is called *counterforce.*

The strategic mind, of course, can always take a war-game scenario a step farther. What if the Soviets should launch an attack on U.S. missile silos but hold back a sizable portion of their own silo-based missiles to deter us from retaliating against Soviet cities by threatening counterretaliation against American cities? To be prepared for this possibility, the Pentagon claims, the United States must have enough nukes to wipe out Soviet silos *at the same time* that we retaliate against Soviet urban industrial targets.

America's nuclear arsenals now house more than 30,000 nuclear weapons: some 22,000 tactical weapons for battlefield use and about 10,000 strategic, or long-range, devices aimed at the enemy's urban-industrial heartland. The Pentagon says these huge numbers are justified because of the threat of Soviet counterretaliation. Critics of the Pentagon believe that the increasing size of our nuclear arsenal means that the United States has been aiming at *counterforce*, or *first-strike capability*, since the 1950s, even though the stated official policy has been one of deterrence. (If our missiles can destroy Soviet silos in a second strike, when the Soviets will be *expecting* action, wouldn't it be more effective to strike first, when we'd have surprise on our side? If *damage limitation* is as effective in a first strike as in a second, doesn't it make sense to go for the first strike?)

In 1967 Secretary of Defense Robert S. McNamara proceeded to order the development of a sophisticated new weapon, the MIRV, or multiple independently targeted reentry vehicle. MIRVs can strike several targets simultaneously, thus overwhelming Soviet antiballistic missiles (ABMs), their means of retaliating against our ICBMs, or silo-based missiles.

We had taken first strike out of the closet and verbally brandished it to discourage lesser challenges. But if we should carry out such a strike, it might insure ultimate escalation instead of squelching nuclear sparring. Once things started, there would be no way we could clamp a lid on. Increasing instability and the rising cost of maintaining our bristling defenses meant that reciprocal arms-control agreements were crucial. In 1972, the heyday of détente, Kissinger and Nixon sought discussion of arms limitation with the USSR. The Strategic Arms Limitation Talks (SALT I) resulted in the first Soviet-American agreement to severely limit antiballistic missile systems, or counterretaliation systems.

The SALT agreements opened what strategists call the Age of Assured Anxiety. As John M. Collins, a defense analyst at the Library of Congress, explains, "U.S. security now depends to unprecedented degrees on cooperation by Soviet competitors, whose incentives to compromise seem less intense than our own." At the same time, strategists thought that in order to deter nuclear war, we had to be seen as being able to fight one. In January 1974 Secretary of Defense James Schlesinger told the press that "in the pursuit of symmetry we cannot allow the Soviets unilaterally to obtain a counterforce option which we ourselves lack." Schlesinger argued for a *limited nuclear response*, meaning that we could launch one or two missiles against selected military targets. A year later Schlesinger announced that the United States would consider first use of nuclear weapons if such a tactic were needed to stop Communist advances in Europe or East Asia. In 1979, for the first time, the North Atlantic Treaty Organization (NATO) decided to place on European soil missiles capable of striking the Soviet Union. Clearly, Schlesinger's *selective targeting* concept had been a way of presenting limited nuclear war as an acceptable option.

There is little reason to believe the Soviet Union would threaten to launch a strategic attack on the United States in the first place, since the Soviets have no guarantee that our response would be limited. Both governments are aware that even if the initial exchanges were limited, they would be likely to escalate. Therefore, as Robert C. Aldridge, a former Lockheed engineer who is now a Pentagon critic, has noted, the only plausible reason for America's developing counterforce capability "is to acquire the capacity to launch an unanswerable first strike against the Soviet Union." Aldridge thinks that in this new race for first-strike capability, America is ahead and should have this ability by the mid-1980s.

Since the early 1970s defense strategists have begun to "think the unthinkable" once again, and to openly talk and write about it. During the 1980 presidential campaign, George Bush discussed the uses of limited nuclear war with a reporter and said he thought we could win one. Carter's Presidential Directive 59 officially endorsed limited war for attacks as well as for retaliation. Herman Kahn, for his part, has been updating his thoughts about the unthinkable. He considers "crazy" the idea that nuclear war is suicidal. In his view **"we don't distinguish between an incredibly unpleasant experience, and a thing which you can survive. If 20 million Americans were killed, there'd be 200 million survivors."**

In 1962, at about the time Kahn first published his ideas, President John F. Kennedy considered inaugurating a government-sponsored program to build fallout shelters. Some people frantically set about digging up their gardens. Companies began manufacturing bomb shelters. Older buildings in some of the country's larger cities still bear the black and yellow signs indicating that they had shelters. But the national program was finally stalled when the respected *New England Journal of Medicine* published a study showing that if a single 20-megaton nuclear bomb were dropped on Boston, only 300,000 people would escape instantaneous and terrible injuries—so few that bomb shelters were obviously futile.

In the last ten years such shelters have been being built again by "survivalists" anticipating economic collapse or nuclear war. These people—no one knows how many there are, for discretion is part of their private strategy—have been putting aside food, energy-producing equipment, and weapons. Some have said they are prepared to kill to protect their stores. Companies that supply survivalists have been doing a brisk business in freeze-dried foods with a ten-year shelf life, fallout shelters (price tag: $8,600), air-filter systems, radiation-proof clothing, radiation meters, anticontamination kits, and other Doomsday paraphernalia. Most survivalists are men; they say women seem to be mainly "desolationists" who would prefer not to survive in a postnuclear world.

People have taken to the streets in Italy, West Germany, France, and Great Britain as part of a new Ban the Bomb movement that calls for "No Euroshimas." Since 1973 Europeans have been worried that the arrival of NATO missiles in their midst means the war-gamers are planning to try out their theories on European soil, that we think limited war is fine so long as it happens to them and not to us. General Charles de Gaulle was determined to develop first-strike capability for France because he doubted there were any circumstances in which an American president would risk the security of American cities on behalf of Europe. A limited nuclear war, Europeans reason, might well be one in which the superpowers fired and then stood back while Europe irradiated. Alexander Haig's statement to Congress in late 1981 that NATO might set off a "demonstration" nuke in Europe as a warning if the Soviets attacked, didn't do much to soothe European concern about the revving up of the superpowers.

IT'S A LONG WAY FROM NAGASAKI

C. Stuart Kelley, research physicist for the Defense Department, was awarded a patent last year for a technical slide rule that calculates damage and other conditions resulting from detonation of a nuclear weapon in battlefield conditions. An operator of the rule adjusts the scale to the yield of a nuclear discharge and can then calculate the blast effects on the environment, the electrical charges in the atmosphere, and the heat output. On its face the slide rule shows the output of a detonation and on its back discloses the effect on military equipment.

This is not a patent destined to gather dust. So far, 1,000 of the slide rules have been distributed to the Army, Navy, Air Force, and nuclear laboratories.

"THE GAP" AND THE ESCALATION OF NUCLEAR POWER

Since World War II, strategists' circular theories of attack and counterattack have been inspired by what they perceive as a series of "gaps" between us and the Russians (see "Us Versus Them"), occurring whenever the Soviets pull ahead in some area of military strength. There was the "bomber gap" in the 1950s, the "missile gap" in the early 1960s, and the "ABM gap" in the late 1960s. Now defense is worried about the "window of vulnerability" that is said to exist because our land-based intercontinental ballistic missiles are vulnerable to a Soviet first strike. (Presumably, in one fell swoop, all 1,052 silo-housed U.S. strategic missiles capable of striking the Soviet Union might be destroyed before the United States could launch a counterattack.) The new MX missile—plus a scheme to keep its whereabouts hidden from the Russians—is an attempt to protect our land-based missiles from wipeout.

In the nuclear age there is no such thing as a "clean weapon," or a "surgical strike" that might knock out U.S. military targets without touching anything else. If the Soviets did try to destroy all 1,052 U.S. silos, blast or fallout would cause the deaths of over half the American population. The Russians would have to assume that after such a calamity, America would surrender. If they misjudged, and we retaliated from SAC planes and submarines, the result would be the destruction of the Soviet Union. (Even if we didn't retaliate, the Soviets would suffer terribly. They would no longer be able to buy American grain, and Soviet crops would be blighted as radioactive dust circled the earth, cooling the atmosphere and depleting the ozone.)

In reality each superpower is still sufficiently vulnerable to make mutual deterrence work. The problems connected with a Soviet first strike are so great it is unlikely that a sane Soviet planner would dare to contemplate one. The gigantic arsenals that have been built since 1949 have brought little real security to either side. Up to 1975 SAC was put on alert 33 times—sometimes by accident.

In *Nuclear Nightmares* Nigel Calder contends that nuclear war might plausibly begin in any of four ways. One would be a showdown between NATO forces in Europe and the Soviets, although Europe is not an ideal theater for "limited" nuclear war. As one NATO officer complained, "German towns are only two kilotons apart." Escalation, in Europe, would be virtually unavoidable.

Calder's second scenario involves a lesser nation with three well-placed bombs launching an attack on another lesser nation that owns only one or two bombs. The third way war could start is by accident, and the fourth is by a cold-blooded—or even maniacal—first strike: hit the enemy's silos before the enemy hits yours.

Both the arms race and the growing threat of nuclear war seem unstoppable because nuclear technology and nuclear politics feed into each other in insidious and dangerous ways. If the political atmosphere between countries is poisonous, one nation is likely to stockpile nukes for a first strike. On the other hand, once one nation owns such weapons, that very fact breeds political distrust. The United States has to take responsibility in large

"Unthinkables" believe that a nuclear war, once begun, is likely to create a disaster of such magnitude that it is not meaningful to plan in terms of actual occurrence.

"Thinkables" believe that one should plan in terms of nuclear war actually occurring, and even for its aftermath.

part for having brought this vicious cycle into being. We were the first country to build nukes, the first and only one to use them against another nation. And we still own the largest number of them.

In thinking that the bomb was the only necessary weapon, the ultimate deterrent, the only solution to war and peace, America made a mistake. The AEC's David Lilienthal said as much. An ardent advocate of domestic nuclear technology (he later opposed the H-bomb and in the early 1960s announced that he wouldn't want to live next to a reactor), Lilienthal argued against single solutions. Realistic, workable solutions, he said, would be more complex.

On the other hand, cold warriors like physicist Edward Teller, who worked on the A-bomb and the H-bomb at Los Alamos, maintain that the nuclear stockpile buys us the time we desperately need to negotiate. **Teller believes that a rational approach consists of having the "courage" to use nukes when necessary, as well as being prepared to survive an all-out nuclear attack.**

HOW TO LEARN TO STOP WORRYING AND LOVE THE ARMS RACE

Considering that all American strategists are concerned about national security, deterrence, and avoiding nuclear disaster, how do they arrive, as do Lilienthal and Teller, at almost diametrically opposite viewpoints? Psychiatrist John E. Mack, who is interested in the psychological effects of the arms race on the way we live day to day, would like to avoid such labels for strategists as "doves" versus "hawks" and "liberals" versus "conservatives." He thinks more apt labels for the two camps are "thinkables" and "unthinkables." Both groups are aware that nuclear war could really happen. But while the "unthinkables," like Lilienthal, consider nuclear war such a disaster that it isn't worthwhile to plan for it, the "thinkables," like Teller, believe we should make plans for both nuclear war and its aftermath.

Mack believes that strategists arrive at such contradictory conclusions because "the unthinkables seem more willing to experience directly, or *hold* emotionally, the reality of nuclear danger, although at times they seem not to appreciate the menace of Soviet political and military strategy as the thinkables perceive it. The thinkables appear unable to experience, or have found a way not to experience, the terror of the nuclear reality itself."

The thinkables' psychological numbness is a useful attitude for a national leader to hold. First, it's politically supportable because it seems like a reasonable reaction to Soviet aggression. And to a country like ours, used to having its own way in international relations, thinkable thinking helps dismiss any lingering twinge of helplessness or powerlessness.

The public, too, has allowed the arms race to continue be-

"It's very important for all of us today to realize that the Soviet Union is not the enemy. Nuclear war is the enemy."—Retired Rear Admiral Eugene La Rocque, Director, Center for Defense Information.

cause it generally prefers to repress nuclear nightmares. One disarmament activist believes that since the 1960s most people have fallen into one of three misconceptions about nuclear war: those who think it might happen but that we could survive it; those who think nuclear war is so devastating that it would never be permitted to happen; and those who think it's a real possibility but that there is nothing to be done about it.

Toy companies have produced scary evidence of the public's psychic numbing: the successful sale of board games for playing at war. In "After the Holocaust" a player competes with three other governments for what's left of North America. In "Ultimatum" you invest in the latest gear to escalate the arms race. In "Containment" players are politicians, antinuke activists, and engineers. Meanwhile, on the video game circuit, one of the most popular items is "Missile Command," which pits three ABM sites, defending six cities, against a barrage of ICBMs, MIRVs, B-1 bombers, killer satellites, and smart bombs. No matter how well the players marshal their defenses, the computer always wins, eventually destroying all six cities. Then the video screen is engulfed in an orange mushroom cloud that flashes THE END.

In the arms race the two superpowers help to create, or increase, each other's power by serving as mirrors for each other. Each side looks in the mirror at its opposite equal, feels scared, and adds another nuke to its arsenal. The arms race means that we all lead double lives: we go about our daily business knowing all the while that the world may end at any time. It would be enormously helpful if ordinary people could find some act that would help them out of this bind. John Mack thinks change will only come about if we acknowledge our terror, "the despair that accompanies the reality of confronting the arms race, and our responsibility for it." Such a shift in thinking would have to be accompanied by a similar shift on the Soviet side, but perhaps a massive change in public attitude on one side could inspire change on the other side, just as fear on one side now inspires more nuke-building on the other.

Titan I missile in a silo in Vandenberg Air Force Base, California. The first talk of a "missile gap" between the United States and USSR occurred in the late 1950s.

chapter 2

Us Versus Them: The Soviet Threat and the Balance of Power

Since 1949, when the Soviet Union first exploded its own atomic device, three short words have dominated American defense planning: the Soviet threat. Anticommunism has always featured in U.S. policy, but in recent years our anxiety and sense of urgency about keeping ahead of the Soviet Union have redoubled.

Until the early 1970s, when the Soviets began to catch up, the United States was always way ahead in the arms race in spite of all the well-publicized "gaps" in defense ability. Often these "gaps" were the result of misleading intelligence reports about Soviet military strength but used expediently by the Pentagon to justify its race to produce new weapons.

The race with Russia took off in the mid-1950s. Americans began hearing frightening stories about a *bomber gap* supposedly produced by the Soviet Union's vastly superior bomber fleet. In response the Pentagon began building its B-52 bomber fleet, projecting that by 1959 the Soviets would have 600 to 700 long-range bombers. By 1961, however, the Soviets had only 190. It was the United States that by that time had built more than 600.

Talk of a *missile gap* began to occur in 1957. In August of that year the Soviets announced their first ICBM flight and a couple of months later launched Sputnik. Intelligence reports trumpeted the fact that the Soviets had 35 ICBMs compared to our 18, and predicted that by 1961 the Soviets would be able to devastate us with no fewer than 100 ICBMs. The presumed missile gap became an enormous campaign issue in the 1960 elections and helped propel John F. Kennedy into office. By 1967 we had built a phenomenal 1,000 ICBMs, far outstripping the Soviets. In 1960 we launched our first Polaris missile. The Soviets didn't have anything comparable until 1968. By that time the United States had enhanced its missile capacity to include the MIRVed Poseidon submarine. The Soviets were unable to deploy a MIRV until 1975.

The reported *ABM gap* of the 1960s again ascribed devastating strength to the Soviets, projecting that they would have 10,000 antiballistic missiles to counter our ICBMs by 1970. Spies reported to the U.S. government that the Soviets were planning to build a huge ABM ring around Moscow. By the time the

Waiting for the Russians. Billions of dollars are spent on equipment for apprehending enemy attack. At this observation center at Fort Monmouth Army Base in New Jersey, guided missiles can be tracked by satellite.

ABM treaty was signed in 1972, the Soviets had actually built only 64 ABMs. Meanwhile, of course, the arms race had undergone another of its characteristic swings between superpower action and reaction: In anticipation of the Soviet ABM system, we developed MIRV (multiple independently targeted reentry vehicles). The Soviets responded by pursuing their own MIRV program on an even more massive scale.

THE ROLE OF DOMESTIC POLITICS IN OUR PERCEPTION OF THE SOVIET THREAT

Over the years U.S. perception of the Soviet threat has veered back and forth between alarm and feelings of relative security. Right now we appear to be in a peak of Russophobia, following a détente period of low threat that ended with the Nixon administration. There's some evidence that U.S. views of the Soviet threat have more to do with what's going on in domestic American politics than with anything the Soviet Union says or does. Nixon noted that he was able to thaw the Cold War temporarily, with his historic trip to China, because he was a Republican; a more liberal Democratic president would have risked charges of being "soft" on communism. (Some observers of foreign policy find it interesting that Russophobia seems to peak during Democratic administrations.)

The first spate of U.S. paranoia about Russia lasted from the end of World War II until the early 1950s and prompted the development of the H-bomb. This peak was followed by the relative calm of the Eisenhower years. Although Ike's administration often foamed with anticommunist rhetoric, the President himself avoided tough-guy confrontations with the Kremlin. He also refused to increase the defense budget for fear of the overweening influence on government and economy of a "military-industrial complex," a term he coined.

The early years of the Kennedy administration followed, bringing a new peak of belligerence toward the Soviet Union. Kennedy increased the defense budget by 15 percent, tripled draft calls, and beefed up civil defense. Unlike Eisenhower, he claimed he would welcome an opportunity to teach the Russians a lesson. By the time of the Cuban missile crisis in 1962, Kennedy had been involved in more foreign policy crises in two years than had Eisenhower in eight.

By 1963, after the embarrassment of the abortive Bay of Pigs invasion, Kennedy initiated the second period of détente, urging a search for ways in which the United States and the Soviet Union might cooperate. The high point of this period of détente occurred during Nixon's administration when he made his trip to China. Nixon also won approval for the SALT I arms limitations treaty of 1972 in the Senate and at the Pentagon. Rather than regard the Soviet Union as bent on unleashing its defense effort to the fullest, forcing the United States to "contain" it, Kissinger and Nixon's détente policies assumed that both superpowers would act as conservative forces in the international political arena. Both statesmen had a great interest in protecting the United States against challenges and instability. But with the Carter administration things changed. Three days after he was elected, in 1976, the Committee on the Present Danger, a group of conservative defense specialists, military planners, and foreign policy experts, held a press conference. **"The principal threat to our nation, to world peace, and to the cause of human freedom is the Soviet drive for world dominance,"** the Committee announced. In its view the Soviets had abused the détente period, using it to fatten defense spending and to obtain general military advantage over the United States. The current crusade to resume the Cold War had begun.

Another group of Cold War enthusiasts, the American Security Council Education Foundation, sponsored a heavily biased television film, *The SALT Syndrome,* which argued against SALT II, the treaty that would have imposed arms limitations on both superpowers. "Such a rapid and intense armament program has not occurred since Hitler's armament of Germany before World War II," the film warned. "We in the United States face greater danger than we did in December 1941 after Pearl Harbor." The film also quoted General Lewis W. Walt, pronouncing, "At the current rate of Soviet military growth, they may soon be in a position to offer the United States an ultimatum to either surrender or be incinerated." The upshot of the debate over how to eval-

WARNINGS OF A SUPERHAWK

"The Soviets have already deployed their first particle beam weapon and the world's largest laser weapon at Saryshagan in the Kazakh Republic, and are testing them. These developments, along with completion of the civil defense system, represent a degree of superiority over the U.S. so imposing that the Soviets will feel free to undertake whatever adventures they have in mind.

"The U.S. has no choice but to begin an urgent national crash program surpassing anything since the Manhattan Project. . . . For the next 20 years the subject of arms control should be wholly ignored because it fulfills the aspirations of only two groups: the liberal press and the Soviet Union."

Major General George Keegan
Statement given to *Discover*, March 1981

(*Note:* In 1977 Keegan resigned as Air Force intelligence chief because he had not been able to convince the CIA and the Defense Department that the USSR was taking the lead in laser and particle beam weapons.)

uate the Soviet position, intensified by the Soviet invasion of Afghanistan, was the defeat of SALT II, which was not ratified by the Senate.

Under Reagan a larger American military budget has been accompanied by Attorney General William French Smith's calls for a stronger Central Intelligence Agency and revitalization of other intelligence agencies. Smith says there are more Soviet spies—disguised as diplomats, trading company representatives, students, scientists, reporters, immigrants, and refugees—entering the United States than ever before, partly because of increased trade, immigration, and exchanges that accompanied détente. Industrial spying, says Smith, has jumped in the microelectronics industry of Silicon Valley. (Microelectronics is an area in which we're definitely ahead of the Russians.) Smith has quoted a Canadian Broadcasting Company report that a theft of inertial guidance technology had improved the accuracy of Soviet ICBMs, making American land-based missiles vulnerable to Soviet attack.

Former Secretary of State Alexander Haig accused the Soviets of using "yellow rain"—fungus-based poison—in Laos and Cambodia. The United States has been trying since 1976 to verify whether the Soviet Union actually has been using chemical weapons against remote villages in Southeast Asia and, more recently, Afghanistan. In 1979 reports of an epidemic of deadly anthrax in the Ural Mountains also prompted accusations that the Soviet Union was violating the biological weapons convention that was signed in 1972 by the USSR, the United States, and more than a hundred other countries.

While the evidence against the Russians remains inconclusive, the issue has enormous political and military implications. We may now face a so-called germ gap. If the Soviet Union has been waging germ warfare, its credibility in other areas of arms control negotiation will certainly come under suspicion. Pressure has already begun mounting in this country for the further development of chemical and biological weapons—perhaps with the help of new recombinant DNA technology.

Further cause for alarm is the question of whether the Soviets have launched or are capable of launching a "killer satellite" designed to hunt down and destroy our communications and surveillance satellites by smashing into them or disabling them with high-powered laser beams. (See chapter 7.) Intelligence reports during the Carter administration estimated in 1980 that the Soviets would have an antisatellite laser weapon in orbit by the mid-1980s. Moscow, for its part apparently alarmed by the military potential of our space shuttle, has implied that it would like to negotiate bans on military activity in the "shoreless cosmic ocean" of outer space.

In general the Soviet Union answers American accusations with claims that the United States exaggerates Soviet military strength in order to advance its own interests and expand its own

"Them"—a couple of Russian radio officers.

military stockpile. Economically and internationally the Soviet Union is not thriving, and its new aggressiveness may be linked to feelings of weakness. The country is in substantial debt to the West. Its agriculture is in such bad shape that it depends on American grain to feed its population. It has indeed increased military spending, in good part for the purpose of defending the Chinese border. The new rapprochement between the United States and China has the USSR feeling that it's surrounded by hostile powers. The war in Afghanistan, which may turn out to be Russia's Vietnam, drags on. Iran and Poland are in turmoil. Some political experts have made the case that since 1960 Soviet influence around the world has actually declined.

At first glance Reagan administration maps and charts of Soviet aggression and military spending seem pretty frightening, but usually they tell only half the story. This is certainly the case with the Pentagon's "Soviet Military Power," a 99-page booklet distributed to senators and congressmen in the summer of 1981 to justify the unprecedented leap in peacetime defense spending. Claiming that Soviet military spending is aimed at "projections of power beyond Soviet boundaries," the booklet manages to mislead in the way that it presents information. It is simply a catalog of Soviet might that supplies no analysis of the relative strengths of the two superpowers. Because of the buildup initiated by the Soviets after the Cuban missile crisis in 1962, they are ahead in numbers of land-based missiles and warheads. However, they are behind in other crucial measurements of military strength. For example, the Russians may have the largest submarine fleet in the world, but their submarines are plagued by ports that are frozen solid for much of the year. The Russians' new SS-20 missile, according to the Pentagon's own description, doesn't seem to be very mobile and is relatively easy for American spy satellites to spot.

Critics of the Pentagon booklet point to at least one good example of the way military bureaucracies on both sides tailor the idea of threat to defend their own policies. An "artist's impression" of the most up-to-date Soviet tank, the T-80, bears no discernible relationship to any existing Soviet tank. It does, however, look a lot like the newest American tank, the M-1. It seems clear that on some occasions, at least, both Pentagon and Soviet planners are guilty of deciding what weapon they want and then inventing threats from the other side to justify the building of it.

This phenomenon is as old as the Cold War itself. In the late 1950s the U.S. Air Force started developing the B-70 bomber. Eisenhower eventually canceled further work on the weapon because it appeared that it wouldn't perform as hoped and was vulnerable to the Soviet air force. In the meantime, however, So-

U.S. Intelligence tries to keep abreast of the state of enemy technology. On display at Fort Meade, Maryland, is some Russian apparatus for use in a relatively crude form of fighting: chemical or germ warfare.

viet air defense planners were reading about the horrifying B-70 in Western technical journals and decided it posed a grave danger to the USSR. Thus, they pushed on with an even mightier Soviet plane, the MiG-25, though in fact the threatening B-70 had actually been dropped.

Needless to say, when the United States heard about the MiG-25, which was reported to be able to fly higher and faster than any other plane in the world, it was eager to pursue the expensive and technologically complex F-15. Then, in 1976, a Soviet defector arrived in Japan in the famous MiG-25, which turned out to be quite unimpressive after all. Although it was supposed to have a top speed of Mach 3.3 (more than three times the speed of sound), pilots were under strict orders, the defector said, never to fly at more than Mach 2.5 because to do so would cause the engine to melt. Nor was the MiG-25 equipped with "look down, shoot down" radar, which detects planes flying at lower altitudes—something the Soviets still haven't been able to perfect.

At the moment the U.S. Air Force is pushing the F-18, which is expected to be the most expensive plane ever built.

The problem with this kind of shadowboxing between military bureaucracies is that we end up with huge weapons programs that are too complex to be properly maintained and that often bear no relation to any real threat from the other side. The money and energy that gets sucked up in developing these projects means that nothing is left over to attend to those military threats that *are* real.

Following are some areas in which the Soviets are gaining ground:

Space war potential At the moment our satellites are smaller, much more efficient, and longer lived. But the Soviets are ahead in laser- and particle-beam technology, and they may perfect a laser weapon powerful enough to knock our satellites out of orbit.

Command and control Reliable command and control systems are crucial. They warn of attacks and enable us to strike back quickly. In recent years, however, our command control systems have demonstrated serious flaws. (See chapter 10.) There have been a number of false alarms. During trials and military exercises computers have failed, so that missiles didn't get fired and troops didn't get the instructions they needed. The Soviets' system may also be flawed and unreliable, but at this writing U.S. defense experts haven't heard about it.

Missiles The Soviets have more than four times as many ICBM warheads as we have missile silos. This is why the United States is currently eager to develop some sort of mobile missile-system. Most Soviet ICBMs are capable of carrying more powerful megaton warheads than ours are. Historically the Kremlin has tended to go for raw power while we have striven for accuracy. Now it appears that the Kremlin is devoting itself to the pursuit of both. To meet the Soviet challenge, the United States is building the MX, which is larger, more lethal, and mounts seven more MIRVs

On guard. In the 1960s a huge underground complex was dug out of Cheyenne Mountain in Colorado. It houses NORAD (North American Air Defense Command), a highly sophisticated warning system that sometimes goes on the fritz. (See chapter 10, "Staging the War with Computers: The World of C^3I.")

per warhead than do Minutemen, our current ICBMs. The MX is expected to be in place by 1986.

The United States is distinctly ahead of the USSR in at least five areas of new technology:

Adaptive optics These compensate for reflection, refraction, and other atmospheric distortions and mean better photography from reconnaissanced satellites and precision targeting for laser and other weapons.
Artificial intelligence The military application of AI is computerized "smart weapons" that can pick out alternatives and make decisions, allowing for more accurate targeting and early warning systems.
Highly energetic munitions weapons These have an explosive power that far exceeds their size and weight.
"Smart" weapons Capable of finding their own targets, smart weapons, once perfected, will change the nature of warfare more radically than did the development of radar during World War II. (See "Radar Review.")
Quiet underwater vehicles At present the Soviets rely on acoustic detection to find our relatively noisy diesel- and nuclear-powered submarines. Quiet vehicles will do a lot to circumvent Soviet surveillance.

The USSR is squarely ahead in two fields of technology:

Controlled thermonuclear fusion Atomic weapons rely on fission—the splitting of atoms—to release their enormous energy. Fusion produces even greater energy by fusing atoms, or forcing their nuclei to join. However, it takes a great deal of energy just to make fusion happen, and neither we nor the Russians have managed to make fusion work efficiently enough to be applied to weapons. Soviet experiments are way ahead of ours on this, though.
Directed energy weapons Specifically, these are the laser and particle beam weapons described in full in chapter 5, and Russia is between three and five times further along than we are in the development of these weapons.

In two other areas the Soviets are challenging us, although we still hold the lead: composite materials of great strength (these will build stronger planes, ships, tanks, and artillery), and Very High Speed and Very Large Scale Integrated Circuitry. The latter constitutes the newest generation of microcomputers. At the moment our software is ahead of Russia's, but the Soviets are making strides in hardware that will decrease computer size, weight, power consumption, and failure rate while increasing computer processing capability. Such tiny computers are crucial to surveillance, weapons guidance systems, and the operation of space satellites.

There's some evidence that emphasis on showy, technically complicated weapons is causing trouble for the Soviets as well as us. Both countries end up with larger and larger blocks of the defense budget supporting a smaller and smaller number of expensive, complicated aircraft carriers, nuclear submarines, and fighter planes—all of which are hard to maintain and repair and are frequently out of operation. Without the test of actual combat, these new weapons are designed to achieve spectacular test performances, not stand up to wear and tear.

"In the face of a determined and capable opponent, the survivability of any spaceborne weapon will depend on short-lived technical advantages over the opponent's system. This survivability will evaporate as improvements are made by the enemy. The confidence in the survivability of one's system will improve again as counter-improvements are implemented to combat the enemy's measures. Thus, spaceborne weapons systems are subject to *technological instability,* i.e., they will be faced with frequent crises of vulnerability that would have to be remedied promptly. The countermeasures and counter-countermeasures cycle promises to be rapid and endless."

Kosta Tsipis
Physicist and Weapons Expert
MIT

In an attempt to create a feeling of security in the world, both the USSR and the United States spend increasing quantities on defense. For some time the USSR has bled the rest of its economy to service and develop its military machine. At the rate our defense spending is increasing, it won't be long before we, too, are purchasing guns at the price of butter.

In a training exercise an infantry soldier under gas attack wears a mask designed to protect against chemical and biological agents.

chapter 3

Binary Nerve Gas: *The "Safer Way"*

One late afternoon in April 1915 a yellowish-green cloud advanced over the fields near Ypres, Belgium. When it had passed, 5,000 Allied soldiers were dead and an additional 10,000 lay wounded, many to become invalids for life.

Some fifty years later, in March 1968, appalled farmers reported that the floor of Utah's Skull Valley, 85 miles southwest of Salt Lake City, was littered with the corpses of more than 6,000 sheep. Apparently the animals had died while grazing.

In both of these instances death was due to poison gas. At Ypres the massacre was deliberate. Indeed, during World War I a total of 1.3 million casualties, including 91,000 deaths, were incurred by chemical weapons. These choked the victims, eroded their lungs, and burned and blistered their skins in a scenario so abhorrent that despite significant advances in chemical technology, there has been no protracted use of lethal gases in combat since 1918.

The Utah sheep kill in 1968 was a different story, a gruesome accident that belied official protestations that such mishaps are "unthinkable." The livestock, it was eventually proved, had been victimized by a lethal cloud of VX nerve gas, which had drifted 30 miles off the open range of Dugway Proving Ground, a sprawling Army reservation, the chief U.S. testing site for biological and chemical warfare.

In 1969 the public outcry resulting from the slaughter at Skull Valley caused Richard Nixon to renounce first use of chemical weapons and to halt their production. At the same time,

Congress imposed restrictions on open-air testing. In 1972 the United States joined the Biological Weapons Convention, which prohibited "developing, producing, stockpiling or otherwise acquiring such agents for non-peace purposes," a stance reinforced in 1975 by America's entry into the Geneva Protocol of 1975, banishing first use of germ or gas armaments. At that time it appeared that the United States had shut the door on chemical warfare as a viable option.

In fact, during the early and mid-1970s the Army Chemical Corps, through research our government described as "defensive" rather than "offensive," continued its study of chemical devices. Due in part to the incident at Skull Valley, emphasis was on a new and supposedly safer breed of armaments—the so-called binary nerve gases that had been on the drawing board since 1949. Today, only a dozen or so years after President Nixon's restrictive order, we are witnessing a startling change of attitude in high places toward the use of chemical devices.

* In a recent statement General Niles Fulwyler of the Army Nuclear and Chemical Directorate announced that . . .

The Department of Defense is presently "developing new doctrine and tactics which will integrate chemical weapons" into the modern battlefield.

* Spending for chemical-war research has zoomed from $29 million in 1976 to $106 million in 1981.
* $23 million has been appropriated by Congress for the building of a binary weapons plant at the Army's Pine Bluff Arsenal in Pine Bluff, Arkansas. Upon completion of the plant in 1984 **the Pentagon plans to begin production of binary artillery shells—the first chemical weapons to be manufactured in the United States since 1969.**

* The Army is currently speeding up training programs at its chemical warfare school at Fort McClellan, Alabama, and concentrating on new antipoison defenses.
* DOD projects an expenditure of $2.47 billion over the next five years for chemical development programs. (Experts outside the Pentagon estimate that these figures could be considerably higher.)

What lies behind this dramatic change in the official position on the creation of a new chemical arsenal?

Perhaps the first event to tilt the scale was the discovery in 1973 of special air-filtration systems in Soviet-supplied Egyptian tanks and personnel carriers. It is possible that these mechanisms were put in place to protect crews from chemically contaminated battlefields. Since the West did not envision chemical aggression, analysts suggested that the presence of such protective equipment could be a sign that the Russians, who also belong to the Geneva Protocol, might be equipping troops to weather an attack of their own making.

Suspicions of foul play increased in 1979, when hundreds of Soviets died mysteriously of anthrax in the town of Sverdlovsk. The U.S. Defense Intelligence Agency attributed the disaster to an explosion in a germ warfare station that caused a lethal cloud of anthrax spores to mix with the town's air. The Soviets, however, insist that the outbreak was caused by the ingestion of tainted meat, animal anthrax being common in the region. The incident has never been conclusively resolved.

This U.S. Army photo, taken in 1943, bears the official caption: "A group of girls at the Johnson and Johnson Mask Company work on the assembly line constructing gas masks."

In the 1980s the climate worsened with reports that it was "highly likely" that the Russians and their sympathizers were allegedly using chemical substances such as the nerve gas GD, or Soman, to flush out Afghan mountain men. In the summer of 1981 the nefarious "yellow rain," a yellow powder purported to have been spread by aircraft over Laos, Cambodia, and Afghanistan and causing nausea, massive internal bleeding, and death, was identified as the chemical T-2 (trichothecene toxin), which is derived from bread molds in the Soviet Union. According to a Jack Anderson column in the fall of 1981, "top-secret intelligence reports" indicate that stocks of T-2 have been stored in Cuba "for possible use in the U.S."

The United States, too, has been accused of waging chemical warfare in recent years through its use of such substances as napalm, white phosphorous, and Agent Orange, in Vietnam. Agent Orange, which simultaneously destroyed crops and revealed enemy hiding places in the jungle, contains the contaminant dioxin, one of the deadliest of poisons. In all, 12 million gallons of the herbicide were used in Vietnam. Veteran groups say that as many as 80,000 U.S. soldiers were exposed to the chemical, but the government claims no link has been established between Agent Orange and such conditions as cancer and next-generation birth defects, which veterans have attributed to the herbicide.

NERVE GAS: THE BRIGHT STAR OF AMERICA'S NEW CHEMICAL ARSENAL

Nerve gases are odorless, colorless liquids delivered in spray or vapor form. They are ten to a hundred times more potent than mustard, and since they cannot be smelled or seen they catch their victims unaware. Their development is rendering phosgene, chlorine, hydrogen cyanide, mustard, and other substances previously packaged in weapon form obsolete.

When absorbed through skin or lungs, nerve gases wreak havoc with the human nervous system by blocking the action of the enzyme cholinesterase, which keeps nerve paths clear by breaking down the chemical messenger acetylcholine. If cholinesterase is prevented from doing its work, acetylcholine builds up in the synapses or junctures between nerve cells, so overloading the receiving cells with messages that the nerve ceases functioning.

As the chemicals take hold a sharp reduction in vision is quickly followed by severe vomiting and diarrhea, convulsions, and eventual paralysis of the respiratory system. Death by asphyxiation can occur in minutes or hours, depending on the degree of exposure. Effects of nerve gas exposure can be cumulative. A sublethal dose, followed within a week or so by a second dose,

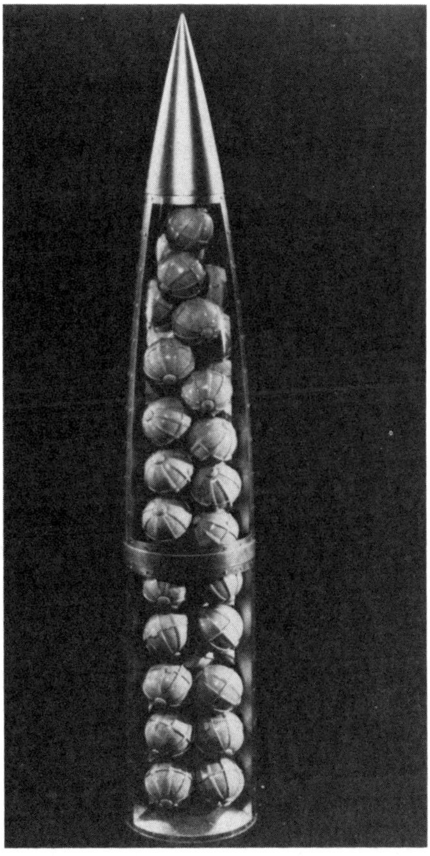

An old-fashioned gas bomb filled with approximately 52 self-dispersing bomblets filled with a gas known as GB, or Sarin. The bomblets spin and glide in a dispersed pattern and detonate on impact. GB kills on inhalation.

can kill. Doses whose effects are not actually lethal often leave psychological and neurological disorders in their wake.

Nerve gases were first discovered by researchers in Germany's I.G. Farben chemical firm in 1936. Laboratory scientists looking for new pesticides stumbled upon Tabun, or GA, an organophosphate poison related to household detergents as well as to bug sprays such as malathion and parathion. During World War II, 12,000 tons of GA were manufactured in a secret factory in the Third Reich, but Hitler, who had been exposed to mustard gas in World War I, apparently never ordered the use of GA. In 1945 the Soviets captured and moved a Tabun-producing plant and a staff of Nazi experts to Russia for research and development purposes.

GA has since become obsolete. Two other so-called G-agents (the Germans stamped gas munitions with the letter *G*) are currently the weapons of choice for the United States and the Soviet Union. The Americans favor GB, or Sarin, which is sprayed in the air and which kills on inhalation. GB is classified as a "nonpersistent" gas since the wind wafts it away within minutes of deployment.

The Russians prefer GD, or Soman, dubbed "persistent" because, unlike the volatile GB, it settles on the ground and retains its lethality. The United States also is working on a persistent nerve gas, VX.

VX is so potent that a quart of it could kill more than a million people and make an additional million violently ill, thus liquidating or incapacitating the population of, say, metropolitan Baltimore.

The consistency of motor oil, VX adheres to all that it touches, penetrating clothing and skin. Even weeks after persistent agents coat metal or glass surfaces of vehicles or weapons, they can kill unprotected troops on contact.

There are two countermeasures to the ravages of nerve gas, one preventive, the other antidotal. Chemical-protection gear, which is standard issue to Warsaw Pact and NATO units, provides an effective physical barrier against nerve gas. (Lieutenant Colonel Harold Shear, commander of the Rocky Mountain Arsenal, quipped, "With a gas mask on, GB can't do a thing for you. I may be able to fly over you at low altitudes and, using its liquid form, drown you in it.")

The medical antidote, atropine, helps to normalize the nervous system that's been attacked by nerve gas, though it does not always succeed in saving lives. In cases of severe exposure, victims might also require hours of artificial respiration in order to survive.

HOW NERVE GAS WEAPONS WORK

Conventional *unitary* nerve gas shells and bombs carry a single deadly payload of active GB or VX and a fairly high risk of

A Department of Defense mock-up of the binary gas "alternative."

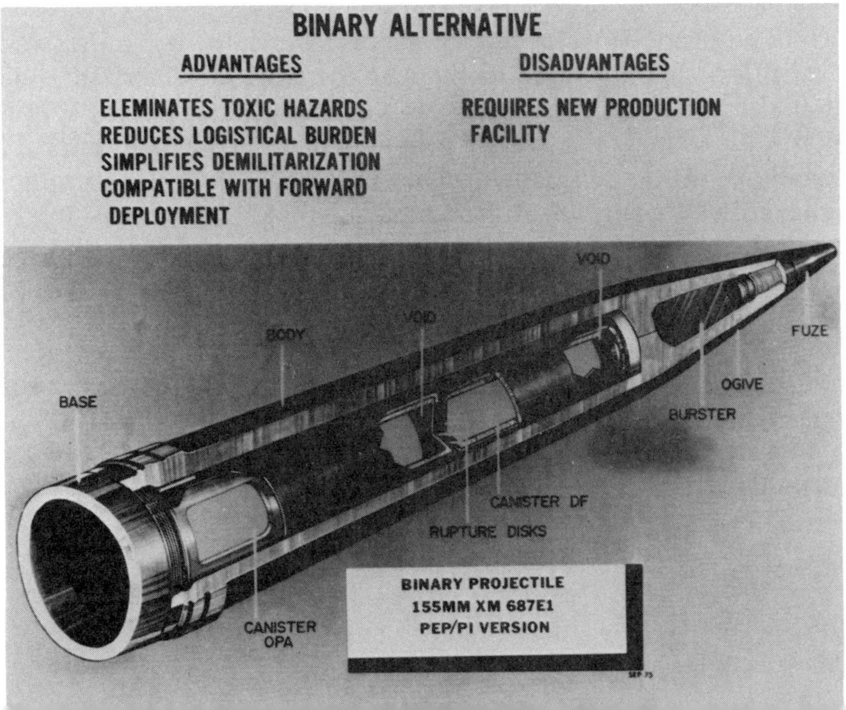

lethal leakage through improper handling. (See the end of this chapter for information on nerve gas leakage in the United States.) The new *binary* nerve gas weapons, on the other hand, contain two substances that mix to form the deadly nerve agent only after the shell, bomb, or rocket has been detonated. The projectile's forward compartment is loaded with a canister of DF (methyl phosphonyl difluoride), which is said to be fairly harmless in isolation. The rear compartment carries a container of isopropanol (rubbing alcohol) plus a catalyst. The isopropanol will not be placed in the shell until immediately before it's fired. After the shell is shot, the metal membranes separating the two parts of its payload burst, and the DF and alcohol mingle. The resultant deadly agent consists of about 70 percent GB.

Present schedules project that production of GB shells at Pine Bluff will begin in 1984.

By 1986 the Army and the Navy will be manufacturing binary weapons delivering VX nerve gas from 8-inch artillery shells and a bomb code-named "Bigeye."

The proponents of binary nerve gas create the impression that this weapon is chemically neat and tidy—a "safer" method than the ghastly unitary nerve gases of the sort used during World War I. But since the recent decision to resume production of nerve gas, some scientists have begun to speak out on the implicit hazards. *Scientific American* reported that DF, the lethal ingredient in the new binary weapon, is about as toxic as strychnine. QL (ethyl 2-[diiospropylamino]-ethylmethyl-phosphonite), one of the weapon's mixers, reacts violently with oxygen and water and will present serious storage problems since it cannot be housed in rubber or in any plastic but Teflon. It also causes respiratory distress, skin eruptions, and gastric upset on contact.

Additional reasons to be concerned about the new binaries have been registered in a report prepared by the Center for Defense Information:

* The component substances of binary agents are commercially obtainable, so that rapid buildup of this new chemical weapon would not require a very complex factory.
* Verification of compliance with a chemical disarmament treaty would be tricky, since facilities for the production and storage of binary chemicals would be difficult to distinguish from commercial chemical plants.
* U.S. production of binaries could legitimate and encourage the spread of chemical weapons. (This view is echoed by Julian Perry Robinson, a British chemical weapons expert, who states: "Once the rest of the world sees the leaders of military fashion getting back into nerve gas, they'll begin to think very seriously about their reasons for not doing so.")

SOVIET VERSUS U.S. CAPABILITY

The *Weekly Surveyor,* a CIA publication, reported that the KGB (Soviet secret police) justified the development of Soviet nerve munitions to the Eastern Europeans as follows: "The U.S.S.R. must maintain a capability in chemical warfare because of the demonstrated capability of the U.S., NATO, and the PRC [People's Republic of China]."

What, in fact, is the size of the Soviet chemical arsenal?

There's a tremendous discrepancy in intelligence reports relating to Soviet stockpiles of poison agents, which Defense Department sources admit cannot be estimated accurately. Figures range from 30,000 to 700,000 tons.

The United States has an estimated 28,000 to 38,000 tons of deadly chemicals, about half mustard and half VX and GB nerve gas. (Nerve gas was first manufactured here during the Korean War. Production reached its height in 1963 and was halted in 1968.)

The Soviet Army has 50,000 to 100,000 troops that specialize in detecting and decontaminating poison gas. The United States has about 6,000 comparable units.

Only defensively might the United States be considered superior to the USSR.

* The M256 poison gas detector kit, an entire chemical laboratory reduced to the size of a stereo cassette, is more compact and more easily workable than its Soviet counterpart.
* The 17-pound M8 gas alarm is thought to be better than comparable Soviet gear.
* A gas mask scheduled for readiness in 1983 will provide greater comfort and peripheral vision than does any model to date.

About one third of the United States stockpile is loaded in approximately 3 million munitions, most of which are stored at Tooele Army Depot, in Utah. (Reportedly the Joint Chiefs of Staff would like to increase that number to 5 million.)

According to General Fulwyler, **1,050 of these devices were leaking at last count.** Leakage of lethal nerve gas substances gained national attention in 1980 when news broke that minute quantities of deadly gas had seeped into the void space of 70 Weteye bomb shipping containers. A cache of Weteyes, GB-filled explosives measuring about 20 inches in diameter and 103 inches in length, are buried about 5.5 miles from Denver's Stapleton In-

In late 1978 members of a U.S. training team demonstrate field-type protective mask and uniform covered with chemical agent stimulant for ground contamination.

ternational Airport under the flight path used by commercial airlines. Although the degree of seepage is 2,000 times less than the acceptable limit per cubic meter as established by DOD and affirmed by the U.S. Public Health Service, Colorado-sponsored legislation states that DOD must detoxify or remove Weteye from the Denver area. The bombs are presently slated to go to the Tooele Depot, to the dismay of many Utahans.

In pushing the new binary weapons, the Army says that our current chemical munitions are obsolete because of modern delivery systems; that the up-to-date binary will deter the Soviets from using similar weapons; and that the binary is safer and cleaner, and thus easier to ship, store, and handle. Because they do not mix and become lethal until fired, binary weapons would be easier to dispose of in case of a chemical disarmament agreement.

Harvard biologist Matthew Meselson, an eloquent spokesperson against the use of the new gases, counters:

* The development of binaries will undercut bilateral negotiations with the Soviets that have been under way in Geneva since 1976.
* Artillery projectiles are not deteriorating significantly and with proper maintenance can remain usable for years. Only a few rockets will become obsolete, and no artillery shells.
* **The new binary weapons will sanitize the concept of nerve gas to the point of making it psychologically acceptable, and thus respectable.**

Further controversy surrounds the question of nerve gas testing. The GB shell that will begin being produced in 1984 got only minimal open field–testing before the 1969 ban. None of the other binaries have been field-tested. The Army claims that laboratory testing with nerve gas simulants is a perfectly adequate way to prepare for this noxious form of warfare, but Saul Hormats, who designed the current unitary munitions, replies: "To embark on a multibillion dollar production program without statistically significant testing would be a criminal waste of money. To supply production line rounds to our forces without confirmatory field trials would be worse."

Preparation for firing a Lance, the missile that carries the neutron bomb.

chapter 4

The Neutron Bomb

Reagan's decision in 1981 to produce and stockpile enhanced radiation or neutron weapons sent shock waves around the world. Encased in an 8-inch artillery shell and delivered in the warhead of a Lance missile (see the end of this chapter), an enhanced radiation or neutron weapon will produce significantly more high-energy neutrons or X rays, or gamma rays, or a combination thereof, than a normal weapon of the same total yield. In effect, it is a miniature H-bomb, fashioned to release energy not as blast or heat but as subatomic radiation that will penetrate armor plate, causing tank crews to collapse within minutes, fatally ill with radiation sickness. In fact it has been considered primarily an antitank weapon, and Reagan's justification for finally deploying it is the Soviets' 4 to 1 tank advantage on the central and northern European fronts. In spite of its incredible lethality, the neutron bomb is considered by some military experts to be "safer" than earlier nuclear weapons because it can be pinpointed or aimed at specific areas, presumably preserving property near or contiguous to the actual field of war. As we shall see, this is a highly controversial position. The fact that its deployment has been conscientiously *avoided* for so long is probably the best indication of just how potentially catastrophic this latest advance in nuclear weaponry actually is.

In concept the neutron bomb is not new. In the late 1940s, soon after the invention of the hydrogen bomb, physicists recognized that a tactical weapon of this kind could be developed. Indeed, throughout the 50s and 60s a small group of scientists, most of whom were working at the government's Lawrence Livermore Laboratory in Livermore, California, perfected the theoretical work on an enhanced radiation weapon. Toward the end of that period a Rand Corporation study praised the neutron bomb for its selectivity and general effectiveness. Actual progress on the bomb was stalled, however, when Secretary of Defense Robert McNamara decided that a European-theater nuclear war would be a losing battle for both sides. Except for developing a Lance missile that would carry a neutron bomb, McNamara deferred spending money on a new generation of tactical nuclear arms.

For the most part this bias against the modernization of tactical devices continued until James Schlesinger became Secretary of Defense in the early 1970s. Although the enthusiasm of the former chairman of the Atomic Energy Commission seemed to flag during the time he was in the Cabinet, he nevertheless ended up allotting funds for the modernization of nuke weapons, stressing the importance of such strategic options as "selective strike." (For a discussion of the history of nuclear strategy, see Chapter 1.)

By 1971 the U.S. government had gone so far as to secure permission for the deployment of a neutron bomb in West Germany. President Gerald Ford put off production of the bomb, however, on the grounds that its manufacture might jeopardize an arms limitation treaty with the Soviet Union.

Adding to the danger of Reagan's deployment decision is the fact that we're not the only ones who have the neutron bomb. The USSR and France have both developed and tested one. In fact, by 1978 France, who withdrew her military forces from the NATO alliance in 1966, announced that she might decide to manufacture a neutron device at some time in the future.

HOW THE BOMB WORKS

The physics of the neutron weapon are fairly straightforward. An enhanced radiation weapon is an H-bomb stripped of its outer shell of Uranium 238. Whereas the A-bomb, which forms the traditional H-bomb's core, is surrounded by a layer of fuel *plus* the uranium coating, the neutron weapon is composed of an A-bomb core and a single layer of fuel. No uranium coating—

that's what makes possible the neutron bomb's unique neutron-shooting capabilities.

> **Step I** The A-bomb, which triggers all H-bomb blasts, is a *fission* device. Fission is the splitting of the nucleus of an atom into two or more fragments. When that happens, the energy that had been used to bind the fragments together is released in the form of kinetic energy, or heat. In constructing an A-bomb, plutonium and uranium are used because both elements consist of unstable, heavy atoms that, when struck by subatomic particles called neutrons, split into two lighter nuclei and release additional neutrons of free energy. These neutrons in turn fission more atoms, creating a rapid *chain reaction* that makes the A-bomb explosion occur. In order for the chain reaction to begin there must be plutonium or uranium of sufficient density to form a *critical mass.* (In the case of the first A-bomb, tested at Alamogordo, New Mexico, in 1945, a lump of plutonium that was below critical mass was compressed into criticality by the detonation of carefully arranged surrounding high-explosive charges.)
>
> **Step II** The enormous heat released by the A-bomb explosion triggers a *fusion* reaction in the layer of fusion material (deuterium and tritium, both forms of hydrogen) that surrounds the plutonium or uranium core. Fission splits nuclei; fusion fuses them. Fusion is the reaction that powers the sun. Under the pressure created by extremely high temperatures, the nuclei of lightweight elements such as deuterium and tritium will fuse together to form helium, at the same time releasing massive amounts of free energy neutrons. These neutrons, notable for their high radiation intensity, are the "bullets" of the powerful neutron bomb. Their energy level is about 14 million electron volts (MeV), which far exceeds the 2 MeV neutrons released by a typical fission reaction. Furthermore, **fusion produces 10 times more neutrons per kiloton of explosive yield than does fission.**

The neutron bomb is a two-part *fission-fusion* process whose most immediate devastation is wrought by radiation. The H-bomb is a three-part *fission-fusion-fission* process. Its final fissioning occurs in the outer layer of uranium (the one missing from the neutron device), and while this second fissioning causes added explosive yield and blast, it also releases neutrons of lower energy than those released by the neutron bomb. (It should be noted here that a nuclear explosion has four effects: *blast* [a shockwave of pressure], *thermal radiation* [heat], *prompt radiation* [largely in the form of neutrons and gamma rays], and *residual radiation* [radioactive fallout].)

Six to ten times more of the energy released by the neutron bomb is in the form of prompt radiation.

THE MORE "HUMANE" BOMB

Neutron devices are hailed by their proponents as being kinder and more humane than the old-fashioned nuclear bombs that wreak their devastation through larger amounts of blast and heat. The argument runs that although those bombs could quite effectively halt Russian tanks, they would in addition devastate appalling numbers of NATO troops, civilians, and property. The accompanying far-flung radiation would amplify the problem, making the occupation and recovery of contiguous affected areas a serious and dangerous matter. In contrast, the high-energy neutron flux of the enhanced radiation device is supposed to provide a cleaner scenario because of its ability to focus tightly on tank-bound soldiers. (This focusing capability has caused Russia to dub the neutron bomb the "ultimate capitalist weapon" since the vaunted advantage of the neutron bomb is the protection of property.)

To evaluate the humaneness of neutron devices, one should know how, in fact, the bombs affect humans.

The ionizing effect of a flux of neutrons colliding with the protons inside the body's cells thickens cell fluid, swells cell nuclei, breaks down chromosomes, and destroys all types of cells, particularly those of the central nervous system.

A further long-term genetic effect is the prohibition of normal cell development due to the delaying—and, sometimes, prevention—of mitosis, or cell division.

The neutron weapon's lethality can be measured most conveniently in rads, the standard measure for radiation doses. Tests the U.S. government ran on rhesus monkeys in the late 1970s caused DOD officials to conclude that "immediate permanent incapacitation" of troops would require 8,000 rads. Since modern tanks have a radiation protection factor of about .5, it would take 16,000 rads to neatly and quickly bump off, or "kill," a tank.

In the last few years NATO has softened its line on the neutralization of invaders, suggesting that perhaps only "immediate transient [rather than permanent] incapacitation" is necessary as a military goal. This would have the happy effect of reducing the amount of radiation needed to do the job to only 2,500 to 3,500 rads (or 5,000 to 7,000 for tanks with the .5 protection factor).

What do all these rads do to people? Following exposure to 8,000 rads it takes a day or two for death to occur, but "incapacitation" is accomplished within five minutes. A dose of 3,000 rads disables its victims with equal speed, but death is delayed for four to six days. Within two hours after exposure to only 650 rads, a human being will become functionally impaired and likely to suffer a lingering, painful process of deterioration leading to death.

At doses significantly lower than the above, radiation does terrible things to human bodies: 10 percent of persons exposed to

150 rads will die of radiation sickness. A high proportion of women, judging by survivors of Hiroshima and Nagasaki, will contract breast cancer. In targets exposed to only 30 rads, defective genes can be expected to show up for ten generations. (Marshall Islanders exposed to no more than 14 rads during U.S. nuclear testing in 1954 subsequently developed leukemia, thyroid nodules, and various assorted cancers.)

Radiation damage from neutron devices will be even greater. Unlike gamma rays, which are thought to have a rad threshold below which no biological damage will occur, neutron radiation is thought to have no threshold. As few as one or two rads could cause cancers and leukemia. Certain biological effects of neutron radiation, such as eye cataracts, leukemia, and genetic damage, are known to be six times more likely to occur than as a result of gamma rays.

The *range* of radiation lethality is also markedly different between fission weapons and neutron weapons. If a 1-kiloton fission weapon were exploded, all those within a 375-meter radius would be exposed to at least 8,000 rads. In the case of a 1-kiloton neutron weapon, the range would incorporate a radius of 850 meters.

One of the advantages of the neutron bomb is the ease with which it can be lobbed from the ground. Here, a tank-mounted howitzer capable of launching a neutron bomb.

Longer-range effects are as follows:

A 1-kiloton fission weapon releases 150 rads out to a distance of .900 meters, 30 rads out to 1.170 meters, and 14 rads out to 1.300 meters.
A 1-kiloton neutron weapon would increase the range of radiation effects dramatically, to 1.7, 2.1, and 2.3 *kilometers.*

THE TOLL ON SURROUNDING CIVILIANS

As Fred M. Kaplan, ex-fellow of the Arms Control Project at MIT, noted in *Scientific American,* the neutron device is far from being a pure fusion device with no blast effect. Indeed, the neutron warhead designed for the Lance missile would release nuclear energy as follows: 40 percent blast, 25 percent thermal radiation, 30 percent prompt radiation, and 5 percent fallout.

Sidney Drell, the distinguished physicist, makes the further point that 1 kiloton of explosive energy at one third of a mile can in fact be highly destructive to property because it creates wind velocities of more than 200 miles an hour. *The neutron weapon does not promise to spare civilians and property to the extent that its supporters envision.*

The particular strategic plan for neutron weapons also makes widespread damage likely. Russian tanks are known to move in two echelons, sometimes three. In nonnuclear situations first-echelon tanks are spaced 75 meters apart; in nuclear scenarios the distance is 100 meters. The second echelon travels about 3 kilometers behind the first. According to defense strategists the first battle of a NATO–Warsaw Pact war would probably be fought in the central region of Europe, where the Warsaw Pact has some 20,000 tanks deployed. Stopping enough first-echelon tanks to induce the Soviets to end the conflict quickly would require the use by NATO of a barrage of hundreds or even thousands of nuclear weapons—probably a combination of low-yield neutron devices and low- and medium-yield fission weapons. As a result significant radioactivity could be induced in the soil, especially if by accident some of the weapons detonated near or on the ground. Large areas would be rendered uninhabitable for long periods of time and the hazard to noncombatants would be grave, especially since the eastern lands of West Germany have become highly urbanized. In such a holocaust scenario Russian nuclear retaliation would be almost certain.

Another consideration is the fact that except for those tankmen fairly close to the detonation, units exposed to the bombs could continue to fight, perhaps more aggressively than before in the knowledge that their own deaths from radiation within days or weeks were inevitable—a kind of nuclear kamikaze. Furthermore, the tanks themselves could continue to be moved ahead since armor-penetrating neutrons would not make the vehicles sufficiently radioactive to prohibit restaffing them with new crews. Thus, ironically, the so-called property-saving bomb will save not only civilian property but also the very weapons against which it is aimed.

Whether or not the neutron weapon will prevent nuclear war is, of course, the most fundamental issue of all. The basic arguments pro and con—or, one might say, the inevitable dilemma it provides—are pinpointed in the Arms Control Impact Analysis of the neutron weapon prepared by the National Security Council in 1977.

> *It can be argued that the improved warhead may make initial use of nuclear weapons in battle seem more credible, which might enhance deterrence. However, by the same token, it can be argued that it increases the likelihood that nuclear weapons would actually be used in combat. In any event, the escalating potential is the same for this weapon as for any other nuclear weapon.*

One last consideration is the cost factor involved in producing neutron weapons—close to $1 million per artillery shell (including projectile, casing, etc.), *and* per the Lance missile that delivers these shells. For the price of two rounds of 8-inch neutron artillery shells, Fred Kaplan notes, the United States could purchase more than 5,000 rounds of conventional artillery shells, three M-60 main-battle tanks, or approximately 50 advanced nonnuclear antitank weapons. The question with which we're left is whether it's in NATO's best interest to acquire an extremely expensive weapon that is likely never to be used, rather than invest in cheaper devices that might significantly improve the West's defensive stance.

LANCE: THE MISSILE THAT CARRIES THE BOMB

A surface-to-surface guided missile, Lance is the weapon best suited for speeding the W-70 Mod 3 (neutron) warhead to its destination. It can also carry conventional high-explosive warheads and has been tested with cluster bomblets. Its estimated range is about 120 kilometers.

Highly mobile, the Lance missile can be taken into battle by helicopter or parachuted from a fixed-wing aircraft. It can swim inland waterways, moves well over ground, and is the first U.S. Army missile to use ready-packaged and storable liquid propellant.

Lance employs a simplified inertial guidance system, one that does not depend on information obtained from outside the missile. (For a fuller explanation of precision guidance, see chapter 8.) Mechanically, in addition to its guidance package, the Lance has a warhead section, fuel tankage, and a two-part Rocketdyne engine. The engine's outer section provides thrust during the missile's boost phase, which lasts about a mile. During boost the missile is closely controlled by the control-and-guidance electronics. When the on-board inertial system detects that the predetermined velocity has been achieved, the boost motor shuts off and the inner (sustainer) motor takes over. At a predetermined point in flight this second motor cuts out, and the missile ends its journey in free flight.

The Lance is presently deployed with several European-based U.S. batallions for training purposes. It has been purchased by Belgium, Great Britain, West Germany, the Netherlands, Israel, and Italy.

SHIVA (named after the Hindu god of creation and destruction) is the largest laser in the world. It has 20 arms (several can be seen here) that deposit their energy—20 to 30 trillion watts of power—to compress and heat the hydrogen gases that make up the fuel in the target chamber.

chapter 5

Lasers and Particle Beams: *Directed Energy Weapons*

High-energy lasers and particle beams, known as directed energy weapons, are characterized by their ability to generate lethal doses of concentrated atomic energy, and train—or pinpoint—that energy on enemy targets. Both particle-beam and laser weapons systems are the focus of mounting controversy. No two studies appear to agree on the time frame, level of technology, or national policy best suited to their development. But in spite of widespread dissension among scientists and Pentagon officials on the practicality of the new directed energy weapons, one thing seems clear. With the fiscal 1982 National Budget's allotment of $195 million to the Pentagon's DARPA (Defense Advanced Research Project Agency) for its program to develop directed energy weapons, this deadly family of arms has advanced to the forefront of the high-tech megaweaponry revolution.

A COMPARISON OF LASER BEAMS AND PARTICLE BEAMS

Both laser and particle beams derive their power from energy unleashed at the subatomic level, but the processes used to release that energy differ. A laser beam is a strongly concentrated flux of coherent or "in phase" light. In the lasing process, units of light called photons, released by artificially stimulated electrons, interact vigorously with surrounding atoms in a dramatic chain-like pattern, producing a highly charged energy pool. This confined energy is further amplified through *oscillation* as photons are oscillated, or sped to and fro between silvered mirrors placed at either end of a tube. While pursuing their headlong course the photons collide with neighboring atoms, releasing energy in the form of additional photons, until the massively energized flux bursts through an unsilvered area, or "window," in one of the mirrors, emerging as a potent beam of laser light. The phrase from which the acronym *laser* derives says it all: *light amplification by stimulated emission of radiation.*

Particle beams, in comparison, can be likened to a flow of bullets. Made up of electrons, or protons, or hydrogen atoms, or ions injected into a particle accelerator from which they emerge in energized streams (see chapter 6), these so-called death rays have the capacity to blast through the atmosphere like the lightning they are said to resemble.

The extreme lethality of particle beams can be traced to two factors: penetration and radiation.

Penetration Unlike laser energy, which must "dwell" on the surface of a target long enough to "burn a hole" in its skin, highly energetic particles penetrate instantly, causing interior devastation.

Radiation As particle beams penetrate they transfer some of their kinetic energy into surrounding matter, generating disabling radiation. The results are burning, melting, collapse of any nuclear material present, and fracture from thermal stress.

> Lieutenant General Kelly H. Burke, Air Force chief of research and development, says of the development of laser weapons for use in space: "If we succeed, we would have a weapon that would change the face of warfare. It would be equal to the invention of gunpowder or atomic explosives."

PARTICLE BEAM: THE NEW "LIGHTNING BOLT" WEAPON

Available information on the particle beam weapon is somewhat sketchy due to the fact that the technology is very new, and the United States is currently engaged in a neck-and-neck development race with the USSR. Indeed, the Russians are apparently working on what one high-level defense official calls "early particle-beam prototype hardware" at their Saryshagan ballistic missile range near the Sino-Soviet border. "Particle beams as weapons are real," says a U.S. beam weapons expert. "Thanks to photographic reconnaissance satellites, we can see the Russian machine taking shape from overhead."

The concept of the particle beam as a weapon dates back to World War II, but the technology for producing such a weapon did not then exist. Today the picture is changing. DARPA and the Navy both support programs of electron-beam development at the government's Lawrence Livermore Laboratory. DARPA and the Army are working on neutral beam propagation at White Horse, DARPA's weapons-testing facility in Los Alamos. These programs have markedly increased the feasibility of "bullet"-wielding particle weapons capable of what DOD calls "catastrophic kill."

Particle beams that are electrically neutral are of particular interest in the development of outer-space megaweaponry. Unlike positively charged protons and negatively charged electron beams, the neutral particle beam is impervious to electromagnetic and electrostatic fields that could interfere with its accuracy and efficiency. According to DARPA spokespersons, space-based neutral-beam battle stations for ballistic missile defense could be ready in the 1990s if sufficient energy levels can be achieved and a system for space-basing developed. The same timetable applies to a fleet of ground-based charged particle devices. (For information on the research being done to beef up those energy levels, see chapter 6.)

What Particle Beam Weapons Can Do

Still in the drawing board stage, particle beam weapons are described by the Defense Department, in a special explanation for the press, as having the following fundamental components:

(1) The beam source, or generator, which consists of a particle accelerator and its associated supply of electrical power, energy storage, and conditioning.
(2) A beam-control aiming and tracking subsystem.
(3) A fire control system.

The goal is to develop a weapon based on a stream of highly energized particles that will travel at almost the speed of light and bore into metal-skinned targets, transferring a large portion of their energy to their objective. As the beam enters the target it would "damage electronic components," and as it continues delivering energy to its objective, would "ignite fuels and explosives and/or create holes in the target."

Special features of the beam weapon include:

Minimal Dwell Time The beam's ability to penetrate the target almost instantly would defeat the enemy tactic of closely spacing its missiles to assure that some, at least, get through.
Durability of fire The "bullets" of a particle beam, being generated by the electrical power input to the beam generator are theoretically unlimited. No running out of rounds of ammunition before the battle's over.
Lethality A chief advantage of the beam weapon is the ease and rapidity with which it can "kill," or destroy, matériel.
Range Experts believe that particle beams may end up having longer range in the atmosphere than laser rays.
All-weather use Unlike lasers, particle beams can penetrate clouds and rain.

Reality or Dream?

The Defense Department stresses the hypothetical nature of present research on particle beam weapons, using phrases like "exploratory development phase," "the establishment of scientific feasibility," and "generic mission development." In spite of certain practical and philosophical reservations, Richard Garwin of Harvard and Kosta Tsipis of MIT are more sanguine. Garwin believes it is "clearly possible" to "eventually" have a space-based particle accelerator capable of "causing destruction on the ground or to high-flying aircraft."

Dr. Tsipis writes and lectures on the many technological difficulties in developing *practical* beam weapons, yet even he agrees: "There is no technological reason to believe" that a machine cannot be developed and constructed some time in the future that would "bore a hole in the outer shell of a missile or a satellite, damage the electronics inside it or explode the high explosives trigger of a nuclear weapon the missile may carry."

LASER WEAPONS

Laser technology has come a long way since the first stream of light photons was lased through a man-made ruby in a lab at Hughes Aircraft in 1960. (Theodore Maiman, who accomplished that task, and Charles Townes, of Bell Labs, are considered the inventors of the laser.) U.S. government laboratories are now at

Miniature "Bell Lab" laser.

work developing test bed hardware to be integrated into a potential system of space-based "killer guns" ready to pump punishing radiant energy into enemy targets.

The key to laser weapons technology is the HE (high energy) factor. In Department of Defense terms, a high-energy laser (HEL) is one that has an average power output of at least 20 kilowatts, (the equivalent of 27 horsepower), or a single pulse energy of at least 30 kilojoules. (A joule equals one watt for one second.) With the invention of the carbon dioxide gas dynamic laser in 1967, the amount of energy required for powering a laser damage weapon system was first demonstrated. (See end of this chapter for specific laser weapon systems the government now has under development.)

What Laser Weapons Can Do

According to Pentagon sources a hypothetical laser weapon system would include the following fundamental parts:

(1) The laser itself, which generates the high-power coherent light.
(2) A beam control subsystem for aiming and focusing the weapon.
(3) A fire control subsystem—the "brains" of the operation—which acquires targets, selects those to engage, tells the beam control system where to look to find them, decides when targets have been destroyed, and designates the next target.

DOD specialists report that the system would function as a weapon that in a high-density threat environment "methodically moves from target to target over its spherical coverage, focuses the beam on the target, holds the selected aimpoint despite the target's speed and maneuver, burns through the target skin, and destroys a vital component or ignites the fuel or warhead. Then, with instructions from its sophisticated fire control system, the weapon switches the beam to the next target providing greatest threat and so continues through tens of successful engagements before the fuel is expended."

Special features of the laser weapon system include:

Speed Since light travels at the speed of 186,000 miles per second, the photon flux would hit its mark almost instantaneously, thereby obviating the need for trying to determine—and aim at—some *future* position of the target.

During the six millionth of a second it takes laser light to travel one mile, a supersonic airplane would advance only slightly more than an eighth of an inch.

Laser Weapon System.
The laser pinpoints a vulnerable spot on the target, such as a fuel tank on a missile, then heats it until it's destroyed. . . . Other possibilities include burning a hole in the skin of a missile large enough to cause massive aerodynamic problems and make it go out of control, or destroying the control electronics by overheating them. . . . Destroying a target with a laser beam is not an easy task. The beam must be focused onto a small spot on a distant moving object and kept there long enough to inflict "fatal" damage—a time that can vary widely among different types of targets. The problems are particularly severe in space, where the beam may have to be focused at a one-meter spot on a target 1,000 kilometers away. —High Technology

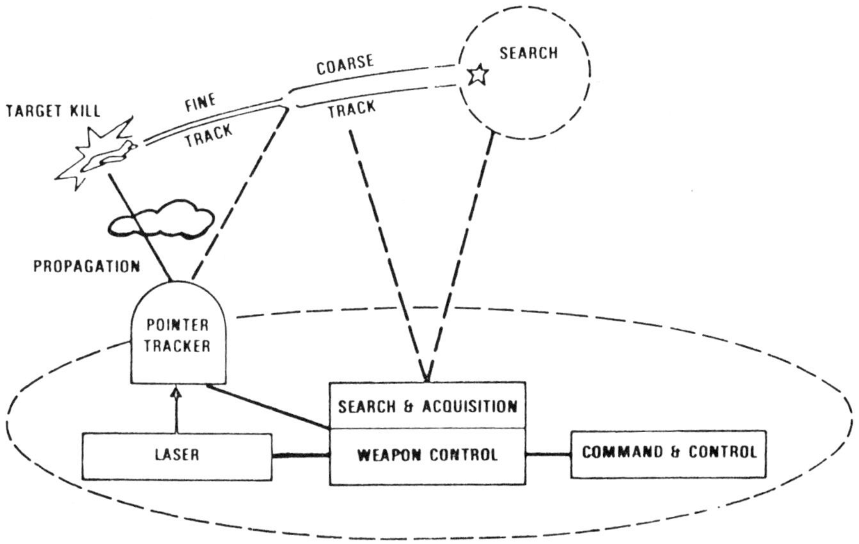

Killing firepower In the area affected, the damaging energy inflicted on targets by HEL killer beams can surpass a nuclear explosion in intensity.

Accuracy Laser weapons would have a high selectivity factor, rendering them capable of pinpointing and destroying enemy targets surrounded by friendly airborne vehicles.

Variation and duration of thrust One laser weapon will be able to engage multiple targets in rapid sequence. Since relatively small amounts of fuel are used to generate the beam needed for each "shot," a single weapon will store a large number of shots.

Range Laser weapons in space would have ranges of thousands of kilometers.

Maneuverability Since its beam is guided by mirrors, the laser weapon has the potential to move rapidly from target to target over a wide field of view.

Problem Areas

Some experts express doubts about the practical application of laser weaponry. In papers presented at the 1981 annual meeting of the American Association for the Advancement of Science (a major segment of the conference was devoted to "Trends in Strategic Weapons and Doctrines and Their Implications for Arms Control"), Dr. Garwin and Dr. Tsipis listed some of the difficulties that would be involved in attempting to deploy these revolutionizing new weapons:

* By deploying metallic foil "parasols," enemy satellites could readily shield themselves from attack by ground-based lasers.
* Adversary satellites, through swift and evasive movement and the strategic use of decoys, could effectively thwart strikes from a high-energy laser.
* Within the atmosphere, high-intensity laser light could cause electrical breakdown of the air, which in turn could disrupt the focusing of the beam.
* HEL fire control systems could be confused by decoys and rendered useless by electronic jamming.

* Low-orbit laser battle stations would be vulnerable to air-launched missiles.
* ICBMs could be protected against laser damage if their surfaces were hardened, thereby necessitating the use of impossibly large doses of photon energy to effect a "kill."

The Department of Defense itself has noted that laser weapons will present certain problems. In order to incapacitate its target, the laser needs not only to strike but to "dwell" on it. There is also the tendency of the atmosphere to absorb the energy of laser beams, causing them to "bloom" or defocus. Inclement weather, especially, affects the ability of the laser beam to get through to its target.

To counteract some of these problems, engineers are racing to perfect finely tuned beam control subsystems with diamond-sharp focusing and targeting capabilities.

In other cases the solution lies simply in placing the weapons in space, which some scientists consider to be the laser's "natural environment." There, free of the degradations caused by the atmosphere, HEL devices can operate at maximum efficiency.

THE SPACE-BASED BATTLE STATION

A system of space-based battle stations is regarded, now, as the most likely near-term scheme for satellite defense. As contemplated by the Defense Department, the scenario calls for putting into orbit big "guns" that will beam lethal rays into the booster mechanisms of hostile missiles, exploding them shortly after lift-off. Undeployed enemy warheads would be destroyed at a blow. "The physics are challenging," says Lockheed's Vice-President of Research and Development, Ray Capiaux, adding, "The Trojan Horse was heroic in its time. The so-called killer satellite could be similarly heroic, with still more profound influences on world history."

Unmanned battle stations, shot into orbit or assembled in space, are not "something out of *Star Wars*," insists Senator Malcolm Wallop of Wyoming, a member of the Senate Select Intelligence Committee and a vigorous advocate of space-based laser weapons. He claims that approximately 36 such devices, each capable of firing 1,000 shots, could cover the globe, with the capability of destroying any missile the Soviet Union might launch. In a study conducted for the U.S. Senate in 1980 by four laser experts, Wallop's estimate was halved. The four consultants, three of them from industry, stated that an effective U.S. ballistic missile defense could be accomplished by putting 18 HEL battle stations in space at an altitude of about 1,100 miles. The station laser devices, measuring from about 19 to 26 feet in length and weighing about 37,400 pounds, would be brought into space by the shuttle. The weapons, positioned in 3 rings of 6 battle stations each, would travel in polar orbit.

Defense officials now estimate that the United States could put beam-wielding devices powered by chemical HELs and sporting complex optical systems into orbit within seven to ten years. Indeed, a killer satellite beam-expander primary mirror is currently under development. (In lasers, mirrors are used to intensify the action of photons of light, thus enlarging or expanding the beam they produce.) The price tag per battle station? From \$2 billion to \$5 billion.

According to one congressional staff member, the 1988–91 delivery date could easily be moved up. "The U.S. energy weapons program is not technology paced," he said, meaning we *have* the technology. "There is no reason why we can't meet the USSR's schedule to fly laser weapons in space by 1984 to 1986."

Senator Wallop concurs, stating: **"High technology is this country's great strength. The United States could beat the Soviet schedule for space laser weaponry."**

TYPES OF LASERS THE MILITARY IS EXPLORING FOR WEAPONS USE

By the mid-1960s several laser types had been developed, but none was suitable for high-energy application. As noted earlier, it was the invention of the carbon dioxide gas dynamic laser (CO_2GDL) in 1967 that riveted the attention of the military.

An Air Force officer conducts a gas laser experiment.

The CO_2GDL Laser

The recognition that gaseous molecular lasers were feasible was a major breakthrough in the development of efficient lasers that generate their energy in the infrared portion of the spectrum. (Color plays an important identifying role in laser technology. The hue of any laser light is determined by the wavelength of its photons. Wavelength also determines the energy level of the photons. The shorter a photon's wavelength, the greater its energy. Thus, low-frequency light such as red light has a long wavelength, and low-energy photons. High frequency light such as violet light has a short wavelength, and high-energy photons.)

The CO_2GDL (laser) works like this: The energy needed to start the lasing process is generated by the combustion of carbon monoxide upon exposure to an oxidizing agent such as nitrous oxide. This explosive union creates excited CO_2 molecules, which retain their excitation level while passing through supersonic nozzles into a lasing chamber. (Nozzle technology is an important area of laser research and development.) As the gaseous flow exits the nozzle, it expands. The excited photons are picked up by mirrors stationed across the energy field, and the oscillation process begins.

A carbon dioxide laser was used in the first successful U.S. laser weapon test, in 1973. With the help of an aiming telescope the laser shot down a winged drone (a pilotless aircraft controlled by radio signals) at California's Sandia Optical Range. A 400-kilowatt CO_2GDL eventually became the laser of choice for the Air Force's Airborne Laser Laboratory (ALL), which is simply a Boeing NKC-135 that has been converted to carry the laser weapon.

Firebee drone planes and submarine-launched Polaris ballistic missiles are the next targets to be blasted by airborne laser weapons.

The Electric Laser

This is a variation of the CO_2GDL in which the initial energy thrust is generated by an electrical discharge rather than by chemical reaction. An electrical CO_2GDL successfully destroyed winged and helicopter drones in 1978 demonstrations put on by the Army. Less powerful than its chemically powered equivalent, the electric laser is favored for use in laboratory experiments that determine laser interaction with materials including titanium, aluminum plate and acrylic plastic.

The Free Electron Laser

Joseph Mangano, assistant director of the government's Office of Directed Energy, has called the free electron laser "one of

This cutout plate of metal demonstrates the precision cutting (i.e. burning) effect of a laser beam.

the more exciting concepts . . . with the capability of revolutionizing chemistry." These remarkably efficient devices combine particle-beam and HEL technology to convert electron beam energy into optical radiation. Here is how the free electron laser works:

A beam of accelerated electrons is passed through a lasing chamber equipped with an alternating magnetic field called a *wiggler.* As the electrons cross through the wiggler magnet they are illuminated by laser light, which they in turn amplify by releasing their own photon energy. Upon exit the devitalized electrons are drawn back into the accelerator, where their remaining energy is recycled for use in accelerating the succeeding electron group.

Free electron lasers promise to have electrical efficiencies of up to 40 percent. In 1982 DARPA planned to spend $6 million researching these remarkable lasers for space-based application.

Rare Gas Excimer Lasers

DARPA is currently investigating potent excimer devices for use in space. Excimer lasers rely on relatively low-powered electron beams to ionize (electrically charge) gas. The ions are then channeled through a chemical process to reach an excited state and to achieve lasing. Excimer lasers typically combine rare gases like argon, fluorine, and krypton.

Both free electron and rare gas excimer lasers operate in the high-frequency range. (Free electron devices can function in the infrared and visible ranges as well.) Of special interest are current military experiments involving the conversion of ultraviolet xenon fluoride into blue-green laser light for use in underwater detection. (Blue-green light can penetrate clouds and water.)

Currently under development is a space-based relay mirror that will target laser beams toward submarines half a hemisphere away.

A laser weapon system that combines ground-based, visible lasers with orbiting relay mirrors has many advantages, in particular:

* Longer "lethal reach."
* Relative ease of deployment in space.
* Unlimited run time.

Advanced Chemical Lasers

Hydrogen Fluoride and Deuterium Fluoride

Fluorine-based lasers have emerged as leading candidates for near-term use as directed energy weapons. According to scientists at the Los Alamos Scientific Laboratory, atomic fluorine has excellent light-producing qualities. Near infrared laser light is generated when atomic fluorine, produced by combustion, is expanded through a battery of supersonic nozzles into an optical cavity where it mixes with hydrogen or deuterium. Combustion within the cavity results in excited hydrogen fluoride, or deuterium fluoride—materials in a ripe condition for lasing.

Programs for the development of weapons based on fluorine lasers include:

* A proposal by Boeing Aerospace Co. to build and test a 2.2-megawatt hydrogen fluoride laser with a 2.5-meter-diameter optical system for use against space-based airborne targets. The device would be placed in orbit by the space shuttle and tested by 1985.
* The use of deuterium fluoride to power a Navy pointer-tracker that in testing has already proved capable of destroying four out of four antitank missiles in flight and of demolishing a tethered helicopter.
* The perfecting by the Navy, as part of its ambitious Sea Lite program, of a 2.2 megawatt mid-infrared advanced chemical laser known as MIRACL. MIRACL is currently reaching the highest energies yet to be demonstrated in the United States, five times those attained by the 400-kilowatt devices used in earlier highly successful tests.

MIRACL is expected to be able to "blind" missiles that are using either heat-seeking infrared sensors or imaging infrared sensors.

"If a missile is looking for IR (infrared), we can supply it with plenty," one Navy official chuckled.

* The development by Bell Aerospace Textron of a 10.6 micron deuterium fluoride/carbon dioxide chemical laser that uses lasing deuterium fluoride to excite carbon monoxide. Experts call this "one of the most efficient lasers ever built."

Oxygen-Iodine Lasers

Another top contender among advanced chemical lasers for weapons is the oxygen-iodine laser, which has shorter wavelengths than fluoride devices. The beam, in the near infrared, is created by exciting iodine with energetic oxygen.

QUASI-SECRET LASER WEAPONS PROGRAMS

COWPONY and COIL

In U.S. Air Force tests currently being carried out at the Los Alamos Scientific Laboratory, under a program code-named COWPONY, the lasing medium is surrounded by chemical explosives that upon detonation produce powerful pulsing effects. As part of the program iodine lasers are also being tested in conjunction with magneto cumulative generators, in an effort to keep up with Soviet research. In these experiments powerful, pulsed laser light is created through the conversion of the chemical energy of explosives into electromagnetic energy.

Under a parallel program, COIL (chemical oxygen-iodine laser), the Air Force Weapons Laboratory at Kirtland Air Force Base, New Mexico, contracted with Bell Aerospace and TRW to develop a 50-kilowatt oxygen-iodine laser device by 1982.

(Chemical lasers are particularly suited for use in space since they don't need electrical energy, can profit from low temperatures to simplify cooling, and can release their highly toxic byproducts with relative safety in the exoatmosphere.)

TALON GOLD and LODE

DARPA is developing subsystems for space-based lasers under the code names TALON GOLD (precision pointing-tracking mechanisms for "acquiring" targets) and LODE (Large Optics Demonstration Experiment). TALON GOLD is using laser radar techniques to achieve the required tracking precision. In 1984 it will use the space shuttle in an experiment in which a low-power laser beam will be pointed to an accuracy of better than 0.2 micro-radians. Currently the ALL (Airborne Laser Laboratory) is equipped with a Hughes Aircraft precision pointer-tracker that uses laser light to acquire and pinpoint targets automatically. The instrument consists of a stabilized platform and an infrared tracking system plus optics that transfer the beam with minimum distortion.

The LODE program is trying to establish the feasibility of large aperture beam control for high-performance space systems. Its aims include the construction of an optical system 4 meters in diameter that can be used with space-based laser weapons. LODE optics demonstrations are scheduled for 1983–85.

A major element of LODE is the development and optical testing of a complex beam-expander mirror to be situated at the interface of the laser cavity and the pointing-tracking system. **Operating much like a telescope in reverse, the device would expand the small coherent beam leaving the laser opening and project it onto a mirror to generate the large-diameter beam needed to travel long distances. Since the beam must be tightly focused on the target, which could be as far away as 1,000 kilometers, the mirror system must be both large and incredibly precise.** (The divergence angle of a beam is inversely proportional to the diameter of the optics and directly proportional to the frequency of the radiation.) The surface of such a mirror must be accurate to within a fraction of a wavelength, and must maintain this accuracy in spite of possible mechanical and/or thermal loads.

In May 1981 United Technologies Research Center confirmed that it had proposed to the Senate's Select Intelligence Committee the development of a low-cost, low-weight beam expander.

The highly polished, silicon-coated mirror, fashioned from thermally stable components, is intended as part of a laser battle station capable of destroying missiles as far as 5,000 miles away.

Officials of United Technologies feel confident that they can fulfill their three-phase Defense Department contracts within four and a half years of go-ahead, researching and developing 2.4-meter- and 10-meter-in-diameter mirror segments at a total cost of approximately $87.5 million.

CURRENT MILITARY APPLICATIONS OF LASER TECHNOLOGY

Rangefinders

XM-1 Laser Rangefinder This 11.3-kilogram device is being developed by Hughes Aircraft for the U.S. Army's XM-1 main battle tank. The time it takes the aluminum garnet-powered laser pulse to reach the target and return is fed into a computer. Almost instantly—within milliseconds—the computer processes the information and generates the required firing commands. The laser rangefinder can distinguish between two targets separated by as little as 15 meters in range.

AN/GVS-5 Laser Rangefinder A portable, handheld model, this neodymium rangefinder weighs 2.27 kilograms. It is about the size of a pair of binoculars.

Antitank Guidance Weapons

Copperhead Cannon–Launched Guided Projectile (CLGP) M712 These antitank missiles, launched from conventional howitzers, home in on laser beams focused by a forward observer. During the first half of the 1980s the Army plans to spend $1 billion on the purchase of 44,000 of these missiles.

Effecting a "Kill."
In one of its first tests at White Sands Missile Range in New Mexico a cannon-launched Copperhead missile homes in on laser energy bounced off the target by a laser designator.

Air-to-Surface Missile Target Designation

Hellfire The Hellfire Modular Missile System, known as Hellfire (from Heliborne-Launched Fire and Forget), is "the U.S. Army's next generation anti-armour weapon," according to the authoritative *Jane's Weapons Systems* published annually in London. The winged, laser-guided missile is not limited to direct line-of-sight attack. It can climb over obstacles, search out targets, and lock onto them automatically. It has proved effective at night and can execute both ripple and rapid fire. The ripple technique involves the launching of missiles fractions of a second apart, at different targets indicated by laser designators on separate codes. Multiple targets can also be engaged if only one designator is used (rapid fire) but the time between launches increases by several seconds.

Between 1978 and 1979 the Army test-launched 32 of the laser-guided Hellfire missiles. There were 30 direct hits.

Laser-guided smart Paveway bombs A smart bomb is a conventional bomb to which a laser-seeker and stabilizer fins have been added. Members of the Paveway bomb family are homed on target as follows: The target is illuminated by laser energy beamed from another source. Light reflected back from the target is detected by a laser semi-active guidance system. (Unlike active guidance systems, which emit the light they will then receive, semi-active systems rely on the interception of light generated elsewhere.) After processing information received, the system changes the bomb's movable surfaces and guides the device toward strike.

Rescue Devices

Laser Augmented Air Rescue System (LAARS) This intriguing mechanism developed by Honeywell permits grounded airmen to establish their whereabouts on a helicopter's imager. A small, handheld retroflector (a mirror designed to direct light back to its source of origin regardless of the angle at which it is received) is exposed to a carbon dioxide scanning laser beam emitted from the helicopter. More complex retroflectors allow difficult-to-intercept verbal communication between air and ground through modulation of the reflected laser energy.

A model of the $5 million Experimental Testing Accelerator at the Lawrence Livermore Laboratory in California. The large structure at the right of the model is the electron gun, which produces the pulse and accelerates it to 2.5 MeV. The central section further accelerates the pulse to 5 MeV. The ETA produces a 10,000 amp electron pulse 40 trillionths of a second long and accelerates it to an energy of 5 million electron volts (MeV).

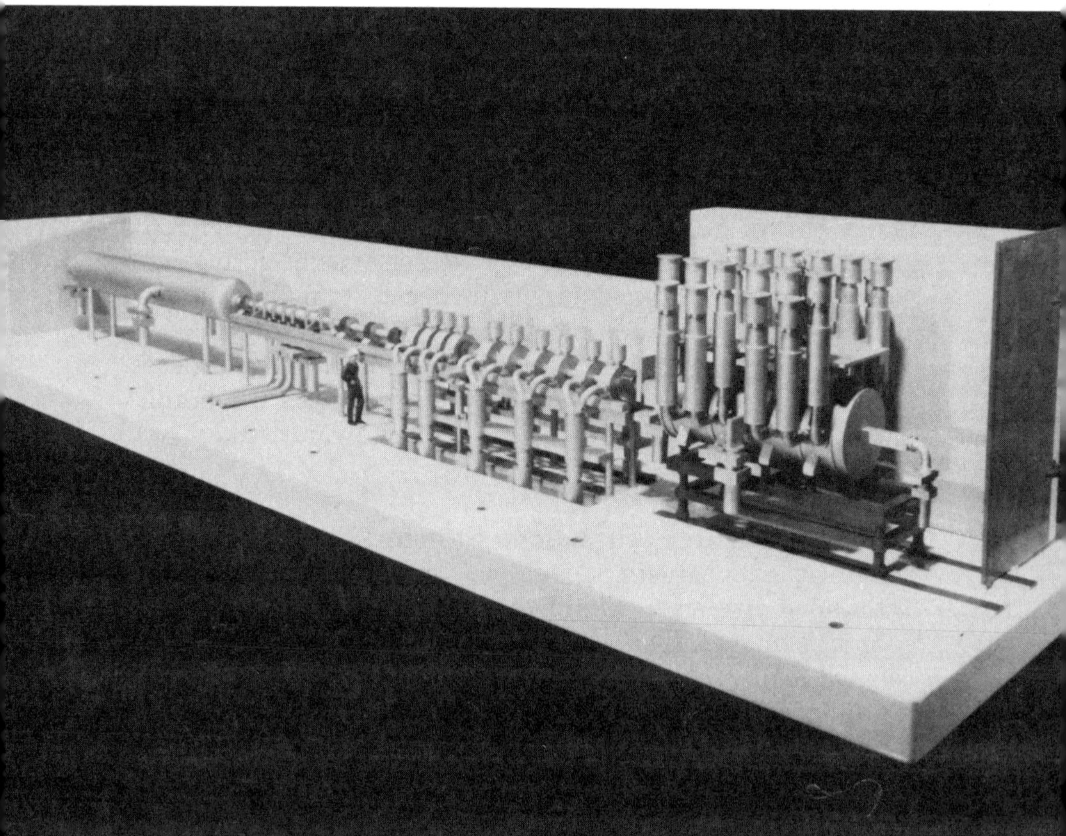

chapter 6

Particle Acceleration: *How the Energy Is Generated for "High Energy" Weapons*

Physicists' understanding of matter and energy has expanded dramatically in recent years. The old model of the atom (which you probably learned in high school), was conceptualized by Ernest Rutherford in 1911 and was based on the idea of the solar system. At the center of the atom is a nucleus, just as the sun is at the center of the solar system. Within the nucleus is located almost all of the mass of the atom, in the form of positively charged particles (protons) and particles that have no charge at all (neutrons). Orbiting about the nucleus, in much the way that the planets orbit the sun, are electrons. These, compared with the nucleus, have almost no mass. Each electron has one negative charge. The number of electrons is always the same as the number of protons, so that the positive and negative charges cancel one another, and the atom as a whole has no charge.

The problem with using the solar system as a model for the atom, modern physicists have discovered, is that the distances between an atomic nucleus and its electrons are enormously greater than we picture the distances between the sun and its planets. As Gary Zukav points out in his overview of the new physics, *The Dancing Wu Li Masters,* "The space occupied by an atom is so huge, compared with the mass of its particles (almost all of which is the nucleus) that the electrons orbiting the nucleus are 'like a few flies in a cathedral' according to Ernest Rutherford."

The following passage from *The Dancing Wu Li Masters* suggests some of the remarkable things physicists have found out in recent years about the reality of the world of subatomic particles.

> *The smallest object that we can see, even under a microscope, contains millions of atoms. To see the atoms in a baseball, we would have to make the baseball the size of the earth. If a baseball were the size of the earth, its atoms would be about the size of grapes. If you can picture the earth as a huge glass ball filled with grapes, that is approximately how a baseball of atoms would look.*
>
> *The step downward from the atomic level takes us to the subatomic level. Here we find the particles that make up the atoms. The difference between the atomic level and the subatomic level is as great as the difference between the atomic level and the world of sticks and rocks. It would be impossible to see the nucleus of an atom the size of a grape. In fact, it would be impossible to see the nucleus of an atom the size of a room. To see the nucleus of an atom, the atom would have to be as high as a fourteen-story building! The nucleus of an atom as high as a fourteen-story building would be about the size of a grain of salt. Since a nuclear particle has about 2,000 times more mass than an electron, the electrons revolving around this nucleus would be about the size of dust particles!*
>
> *The dome of Saint Peter's basilica in the Vatican has a diameter of about fourteen stories.*

Imagine a grain of salt in the middle of the dome of Saint Peter's with a few dust particles revolving around it at the outer edges of the dome. This gives us the scale of subatomic particles.

> *It is in this realm, the subatomic realm, that Newtonian physics [the law of gravity, the laws of motion, etc.] have proven inadequate, and that quantum mechanics is required to explain particle behavior.*
>
> *A subatomic particle is not a "particle" like a dust particle. There is more than a difference in size between a dust particle and a subatomic particle. A dust particle is a* thing, *an object. A subatomic particle cannot be pictured as a thing. Therefore, we must abandon the idea of a subatomic particle as an object. . . .*
>
> *At the subatomic level, mass and energy change unceasingly into each other. Particle physicists are so familiar with the phenomena of mass becoming energy and energy becoming mass that they routinely measure the mass of particles in energy units. (Strictly speaking, mass, according to Einstein's special theory of relativity,* is *energy and energy* is *mass. Where there is one, there is the other.)*

Physicists are learning more all the time about the nature of subatomic particles; yet these particles have already been commandeered into performing some of the century's most awesome tasks, ranging from fissioning atomic bombs to serving as the lethal "bullets" of the new directed energy weapons. **The harnessing and amplifying of particle power, whether to explore the nature of matter or to demolish enemy missiles, is the work of a remarkable twentieth-century machine: the particle accelerator.**

HOW THE PARTICLE ACCELERATOR WORKS

A particle accelerator is a device capable of trapping electrically charged particles in powerful electromagnetic fields and accelerating them to very high speeds. Inside the accelerator the particles are sent smashing into one another so that researchers can find out what the particles are made of. The particle that does the smashing is called the projectile, and the particle that gets smashed is called the target.

The most advanced particle accelerators send both the projectile and the target particles flying toward a common collision point, which is located in a device called a bubble chamber. It is here, in the bubble chamber, that the behavior of smashed particles—which leave trails behind them like the trails of jetliners—is closely observed. Each time a particle enters the bubble chamber, a computerized camera takes a photograph. All particles are too small to observe directly, and most live for less than a millionth of a second. Therefore the only way to observe them is to catch their trails on the photographic plate. (Located inside a magnetic field, the bubble chamber causes positively charged particles to curve in one direction and negatively charged particles to curve in the other. The mass of a particle can be determined by the tightness of the curve it makes. Lighter particles curve more than heavier ones.)

The electron volt (eV) is the standard unit of measurement for particle energy. During the last fifty years achievable voltage in particle accelerators has soared. Early accelerators from the 1930s—the Van de Graaff and the Cockcroft-Walton—were unable to reach energy levels above a few megaelectron volts (MeVs). The new Tetravon synchrotron accelerator (see section to follow on cyclotrons and synchrotrons) being constructed now at the Fermi National Accelerator Laboratory (Fermilab) in the middle of the Illinois prairie *will function in the tetraelectron volt range.*

Note: A *kiloelectron* (KeV) is a thousand electron volts.
A *megaelectron* (MeV) is a million electron volts.
A *gigaelectron* (GeV) is a billion electron volts.
A *tetraelectron* (TeV) is a trillion electron volts.

Voltage defines the velocity and energy with which the subatomic "bullets" in a beam weapon strike.

THE LINAC (ENERGY-BY-THE-YARD)

Scientists eventually recognized that in order to achieve higher energies, it would be necessary to accelerate particles in stages, giving them a series of small boosts rather than a single big one. The most obvious way to do this is to set up a row of accelerator stations within a device called a *linear accelerator,* or linac (*lin*ear *ac*celerator). Particles in linacs are in effect accelerated by the action of an electromagnetic wave traveling down the tube. Particles traveling at nearly the speed of light "ride" the wave like surfers.

The Stanford Linear Accelerator, dubbed "The Monster" by the men who work with it, is designed according to this principle. Located at Stanford University in Palo Alto, California, The Monster was completed in 1961 at a cost of $115 million. It is 2 miles long and contains 82,560 radio frequency cavities, or energy way stations. Designed to raise electrons to 22 GeV in its 4-inch-wide copper accelerating pipe, it is being perfected to accelerate particles to 30 GeVs or more.

The most powerful electron accelerator in the world, the Stanford linac provides unparalleled precision in the measurement of the smallest atomic particles. The internal structure of elementary particles can be seen down to 10^{-15} centimeters.

CYCLOTRONS AND SYNCHROTRONS (ENERGY-IN-THE-ROUND)

As part of the search for higher particle-energy production, the cyclotron was pioneered in 1932 by Ernest Lawrence at the California Institute of Technology. This revolutionary device operated on the theory that the same accelerating station could be successively used to feed energy cumulatively into orbiting charged particles. Atomic "bullets," accelerated in an expanding spiral to very high velocities, were held on track by strong electromagnetic fields. The advanced Lawrence cyclotron set up at the University of California in Berkeley in 1939 was one of the world's first big atom smashers. (It was dismantled in 1962, having become obsolete.)

The concept of energy creation "in the round" was later refined into a fixed radius machine called a synchrotron. **Synchrotrons generate the highest energies yet obtainable by propelling particles around a circular track equipped with multiple acceleration stations and a series of magnets that keep the beam controlled.** They are called synchrotrons because the particles synchronize their motions automatically with the rising frequency of the accelerating voltage and the rising magnetic field.

The imposing Brookhaven Alternating Gradient Synchrotron (AGS), completed in 1960 in Upton, Long Island, achieves energies of 33 GeV. Within its main accelerator ring, which measures 843 feet in diameter, orbiting particle clusters are exposed to magnets that continually squeeze and shove them with the changing magnetic fields or gradients that give the AGS its name.

ENERGY THROUGH COMBINATION

Scientists have found that one way to soup up particle power is to incorporate different acceleration techniques into a single system. This can boost energies by a factor of up to 100 at each stage of the road. Along the way, yesterday's superstar devices become reduced to stepping-stones in a larger process.

In the Brookhaven AGS, for example, a Cockcroft-Walton generator, originally an accelerator in its own right, provides the initial electric charge that bounces protons into a linac. There, particles are accelerated to energies of 50 MeVs and speeds of 62,000 miles per second—about a third of their eventual speed. At Brookhaven the final step is the main (circular) acceleration chamber.

There, protons make 300,000 orbits, or a distance of 150,000 miles in about one second.

In other accelerator systems, however, particles could continue to pick up more energy by moving into a second acceleration ring. This will be the case with the Fermilab Tetravon, which will use superconducting magnets twice the strength of those in current use. The Tetravon will produce a beam with the total energy of 8 million joules—about the energy of a 100-pound artillery shell after leaving the muzzle.

Acceleration systems can attain even greater "combinatorial richness" through the inclusion of storage rings that keep particles revved up and ready for action.

WEAPONS APPLICATIONS

Particle beam weapon technology is essentially the same that's used in particle physics research. *The only striking difference is the addition of aiming and fire control subsystems for weaponry application.* Defense Department prototypes for beam weapons rely on linear accelerator technique. The same is true of the Soviet beam weapon currently under construction at Saryshagan, in southern Russia. Indeed, a major part of the funding currently allotted by DARPA to its particle beam program is slated for accelerator development.

The DOD program has two major thrusts: development of charged particle (electron) beam accelerators suitable for within-atmosphere application, and the creation of accelerator systems capable of producing electrically neutral beams to be used in space.

It is part of the government's plan that beam weapon accelerators be scaled to compact sizes consistent with spacecraft and aircraft use, while retaining the hundreds of MeVS necessary to insure "kill."

IN-ATMOSPHERE APPLICATIONS: THE ETA AND ATA

The government's Lawrence Livermore Laboratory has already generated electron beams with its linear ETA (Experimental Test Accelerator). After acceleration, beams are transported through a 2-meter-long vacuum chamber filled with low-pressure gas for observation. Beams will exit the ATA's particle injector at 2.5 MeV and a velocity of 0.985 the speed of light. They will then accelerate to about 0.999 the speed of light and 50 MeV energies. The major effect during acceleration will be a significant increase in electron mass. The government expects ATA beam propagation experiments to begin by late 1983 and be carried out over kilometer distances. The experiments are intended to determine what happens to beams between end of acceleration and strike.

SPACE-BASED WEAPONS APPLICATIONS: OPERATION WHITE HORSE

Electrically neutral particle beams—immune to both electromagnetic and electrostatic fields—are the beams best suited to use in space. The Army's neutral beam technology program, code-named White Horse, is expected to follow the testing and demonstration procedures employed for electron beam research at the Lawrence Livermore Laboratory. A 5 MeV accelerator test stand at Los Alamos will be followed up by an advanced model operating in the 50 to 100 MeV range.

The provisional timetable calls for 5 MeV neutral beam experiments beginning in late 1983, and advanced accelerator testing four years later. Successful results with the more powerful device, according to the Defense Department, could lead to the initiation of studies for optical systems designs for weaponry application. If energies could be scaled to 500 MeV (kill) levels,

---

A space-based neutral beam antisatellite weapon could be ready by the mid-1990s.

Columbia, *the first U.S. space shuttle. The Department of Defense expects to launch all its military spy satellites on the shuttle, using NASA tracking facilities and astronauts. Former Secretary of Defense Harold Brown has said, "By the 1980s we will be almost totally dependent on the shuttle for our national security space missions."*

chapter 7

Shuttles and Killer Satellites: *Warfare Moves into Space*

The picture-perfect touchdown of the orbiter *Columbia* in the California desert on April 14, 1981, reestablished the United States as the world's leading space power. Perhaps even more momentous was the shuttle's second return to earth in November of that year, proof that a reusable space vehicle was more than a National Aeronautics and Space Administration (NASA) designer's dream and affirmation that America had not lost its technological touch.

A great deal was riding with *Columbia* and her crew on those two historic flights. As John Noble Wilford, science correspondent for *The New York Times,* wrote shortly before the first lift-off, the shuttle's success in this time of national doubt would serve to remind us that "we are not quite the paralyzed giant of our darkest self-image." A visible space shuttle would also foster our traditional commitment to adventure and frontiersmanship, he continued, facilitating our role as a "spacefaring people."

TECHNO-RUNDOWN

The space shuttle, a transportation system designed to haul orbital freight, consists of three main components with a combined lift-off weight of 4.5 million pounds: a delta-winged orbiter (the first of which is called *Columbia*); a huge, bullet-shaped fuel tank on which the orbiter rides into near-orbit; and two solid-fuel rocket boosters that add thrust to the orbiter's three main engines during the first few moments of ascent.

Unlike anything that has flown before, *Columbia* is designed to operate like a spacecraft while in orbit and return to earth like a mammoth glider executing a runway landing. Within only two minutes of lift-off the orbiter reaches a height approximately 28 miles above earth. At this point the boosters, each having consumed its 1 million pounds of fuel, are parachuted into the sea. (Upon retrieval they are refurbished for future use.) The 154-feet-long, 27.5-feet-in-diameter fuel tank continues mounting skyward with the *Columbia* astride. About eight minutes after launch, just before the shuttle goes into orbit, the tank, emptied of the supply of liquid oxygen and hydrogen with which it has fueled the *Columbia*'s principal engines, is jettisoned. It's the only element of the shuttle that's expendable, and whatever portion doesn't burn up when entering the atmosphere rains down in chunks into the Indian Ocean.

The orbiter itself is a 122-feet-long, 150,000-pound spacecraft that is directed through space by an Orbital Maneuvering Subsystem (OMS) comprising two small rockets. These fire several times per mission. Capable of carrying payloads weighing up to 65,000 pounds into low earth orbit in its 65 feet-by-15-feet cargo bay, the spacecraft requires a crew of two and can transport up to five additional passengers. The first 50 shuttle flights (through 1985) are almost sold out. Missions typically last from four to seven days but could extend to a month. A fleet of 32 shuttles is currently projected.

The *Columbia* is equipped with six computers, four identically programmed, the others serving as backup machines. Interestingly, these devices, each of which consists of two 55-pound boxes, one to handle data flow and the other to compute, are not the latest models but modified versions of equipment developed in the early 1970s by IBM for use aboard military aircraft. The use of multiple computers that continually check one another for errors is called *fault tolerant computing,* a technique that uses redundancy to enhance reliability.

Virtually every step of the shuttle's mission can be handled by computer except for lowering the landing gear and braking on the runway, tasks the astronauts must perform manually. At all other times the astronauts' instructions are processed through the computers rather than going directly to the shuttle's maneuvering devices. Indeed, as noted by Gentry Lee of the Jet Propulsion Laboratory in Pasadena, California, "The software [computer programs] essentially run the mission."

Computers following instructions punched in by the astronauts initiate the hour-long landing process by firing 44 small thruster rockets, which turn the shuttle 180 degrees so that it is moving backward. They next fire the two OMS rockets for about two and a half minutes in what is called a "deorb burn." This slows the shuttle sufficiently for it to fall out of orbit. The astronauts then refire the thrusters, turning the shuttle forward again, and the computers go into their entry guidance mode, angling the orbiter by means of the thrusters so that its nose is lifted about 40 degrees when it hits the upper atmosphere at 400,000 feet. This angling causes the brunt of the frictional heat of reentry to be borne by the undersides of the fusilage and wings. These are covered with a protective coating of more than 30,000 heat-resistant ceramic tiles. During the shuttle's first few flights instructions for deorbiting and trajectory adjustments are being initiated by Mission Control at the Lyndon B. Johnson Space Center in Houston. Thereafter, the computers will plan descents on their own.

Once within the atmosphere, the shuttle becomes an airplane—or, more precisely, a glider—soaring at supersonic speed, its wing flaps and rudder computer-controlled. The astronauts take over the guidance process about 20 minutes before landing, to bring the craft to touchdown.

PEACETIME USES OF THE SHUTTLE

On return from the *Columbia*'s first voyage, astronaut John Young declared, "The American public is going to get their money's worth out of this baby." In addition to its military uses, which were played down by NASA in the beginning (the shuttle as weapon system will be discussed shortly), the shuttle is expected to foster advances in a number of fields, including:

Communications The shuttle will help meet the demand for satellite communications by putting switching stations into orbit. (Satellite messages are currently switched on the ground, but as the number of communication satellites increases, orbital "real estate" above earth-based switching stations is running out.)

Pharmacology In a condition of microgravity, or weightlessness, such as exists in space, it is easier to separate the components of liquids and keep them apart. Thus, scientists foresee the manufacture aboard the shuttle's Spacelab of substances of near-perfect purity, particularly hormones and vaccines.

Metallurgy Again, due to microgravity, substances that will not mix on earth can be combined in space. This should permit the creation within Spacelab of new alloys capable of replacing scarce materials.

Space Colonization By transporting workers and materials to building sites, aided by smaller load-bearing shuttlecraft called space tugs, the shuttle would permit the colonization of space. By helping to create orbiting space settlements, the shuttle may ultimately insure the continuation of the human race in case of lethal earth pollution or nuclear holocaust.

Deployment of Space Observatories Notable in this area would be the placing in orbit of NASA's Space Telescope sometime in the mid-1980s. This 10-ton 43-feet-long and 14- feet-in-diameter unmanned observatory will have a 96-inch-in-diameter primary mirror measuring only a few inches less across than that of the Mount Wilson telescope in California—long the world's largest. The Space Telescope should increase 350-fold the visible volume of space now accessible with the finest ground-based telescopes, since in space there is no atmospheric attenuation between the telescope and the distant object. It will record objects up to 7 billion light-years away with the same clarity with which present earth-based telescopes now capture objects at a 1 billion light-years' distance. Astronauts will service the telescope and collect film for processing.

Earth Science Carrying instruments that execute a more discriminating and effective "remote sensing" of our planet, the shuttle will help cut through the deluge of data accumulated previously by satellites. Shuttle-based experiments like FILE (Feature Identification and Location Experiment) will measure landscape brightness through simultaneous readings in both visual red and infrared light, comparing the two in an attempt to characterize such features as water, crops, and ore-bearing rocks.

MILITARY USES OF THE SHUTTLE

Perhaps the most controversial aspect of the shuttle is its tie to the military. The shuttle program, which emerged in the wake of Apollo, was a compromise from the start (NASA had wanted a 100 percent reusable craft), and even after authorization by President Nixon in 1972, it might not have survived mounting costs without the decision to take on payloads for the Defense Department. Says Dr. George Mueller, who left NASA in late 1969, "Without the military in, there wasn't much long-range hope." And on the eve of the *Columbia*'s first launch, NASA's Director of Space Transportation, John Yardley, remarked, "We did need some money. We did need the support of the Department of Defense."

The Defense Department booked a third of the shuttle payload space back in 1973, and most of its cargo is classified. Although NASA repeatedly stresses that only 30 percent of projected missions are DOD-related, **some U.S. officials are beginning to admit that the military has assumed a dominant role in the shuttle program.** Dr. Hans Mark noted in 1980, while he was Secretary of the Air Force, that "NASA is in fact a minor user [of the shuttle] and not the driver. That's something the NASA folks don't like to hear, but it's true."

Military functions of the shuttle include the following:

Use as a launching platform for spy communication and navigation satellites, according to Pierre Kohler of the Meudon Observatory outside Paris, who estimates "about one-half of all existing artificial satellites are spies." Many of the satellites would be sent 22,300 miles above earth into geosynchronous (stationary) orbit by *Columbia,* a height that the shuttle itself cannot attain. The satellites would be borne out of low earth orbit at the tip of one-way rockets called inertial upper stage (IUS), presently being developed by DOD. At launch time the rockets would pop up on their cradles like tiny Titan rockets and be propelled by springs out of the shuttle's payload bay into space. They would be fired some 45 minutes later, when well clear of the orbiter. Launch into geosynchronous orbit via IUS rockets is notably cheaper than launch by surface-based Titans: $18 million per shot for the IUS and its launcher as compared to over $100 million per ground-based missile.

Service as a laboratory for research on satellite laser weapons and a workshop for their assemblage and launch. (For more about laser battle stations designed to detect oncoming enemy ICBMs and destroy them with beams of lethal light, see chapter 5.)

Employment as the United States' key defense against Soviet killer satellites (see page 95), which in turn are being developed by the USSR to protect against the shuttle, an irony all too characteristic of arms-race scenarios. Already the Soviets are worried about the 50-foot articulated manipulator arm that the shuttle will bear aloft. Equipped with wrist, elbow, and shoulder joints, the mechanical limb is planned to scoop up U.S. satellites for purposes of repair or return to earth, an extremely delicate operation since an inopportune nudge could send a satellite spinning away through space. Although there has been no talk of garnering enemy satellites, a major military impetus for the shuttle is said to be its potential to remove killer satellites from space.

Service as a surveillance vehicle from which to inspect enemy satellites. The shuttle could also be used as a construction site for a huge space-based surveillance platform and for manned orbital posts serving early warning vehicles.

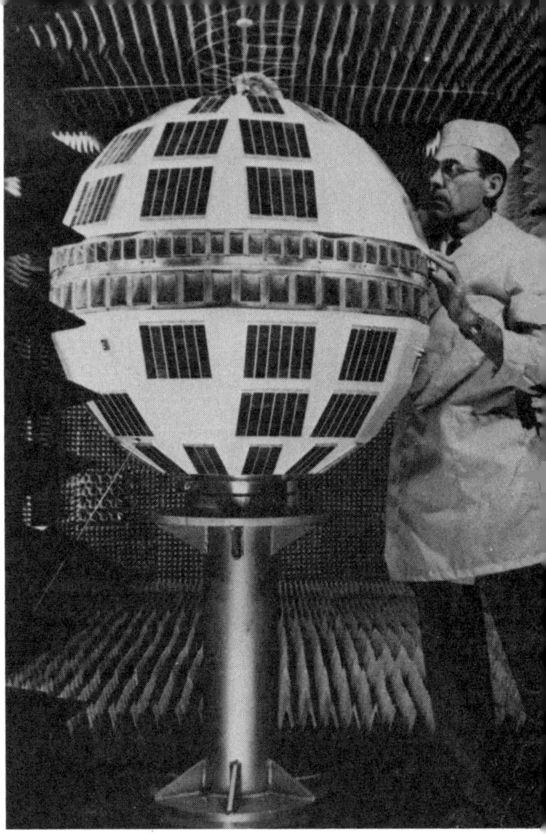

Telstar, the first satellite, underwent repeated tests before its successful launch in 1962. Here, a Bell Labs engineer inspects a model of the satellite in a room lined with pyramids of foam plastic that absorb radio energy. The chamber simulated the radio environment of space so engineers could test the satellite's antennas.

Any doubts about military involvement with the shuttle have been dispelled by the DOD space-shuttle launching complex presently under construction at Vandenberg Air Force Base in California. It will be ready by 1985. Since launchings from Florida's Cape Canaveral are better placed to perform surveillance and communications functions requiring *equatorial orbits*, the $2 million West Coast facility will be used to place military reconnaissance satellites into *polar orbits* that can insure space-based inspection of virtually the entire world.

KILLER SATELLITES: THE RUSSIAN RESPONSE

The Soviets have been pioneering space weapons in the form of antisatellite interceptors since 1968. During the first ten years of experimentation they launched 15 of these trial "killer satellites."

In 1977, Cosmos 970, the sixteenth in the Soviet series, was fired, establishing Russian supremacy in the area of spaceborne satellite interception. After being put into low orbit, Cosmos was sent to an altitude of 620 miles. In less than two orbits it had worked its way to another Soviet satellite beside which it circled the globe four times before veering away. This feat caused former Secretary of Defense Harold Brown to acknowledge publicly that the Soviets had developed a limited "hunter satellite" system.

The Soviets have held onto their lead in orbital weapons. Presently considered ten years ahead of the United States in this domain, they have continued to research and develop the hunter satellite, with mixed results. Their perceived aim has been the perfecting of an infrared sensing device that would permit killer satellites to home in on heat radiated by space-based targets.

In fact, in the first known Soviet operational test since the launching of Cosmos in February 1981, infrared homing was not employed. Nevertheless the trial provided new proof of the Russian ability to position vehicles in space with great accuracy. According to intelligence sources a weapon was orbited around the earth, then maneuvered close enough to a target satellite to demolish it, although destruction was not attempted. A month later, in March 1981, an experimental Russian hunter satellite equipped with a radar homing device reportedly overtook a space-based target satellite. It then blew itself up with a nonnuclear device that disintegrated into pellets and shrapnel. Although the target satellite was not destroyed by the blast, had it been carry-

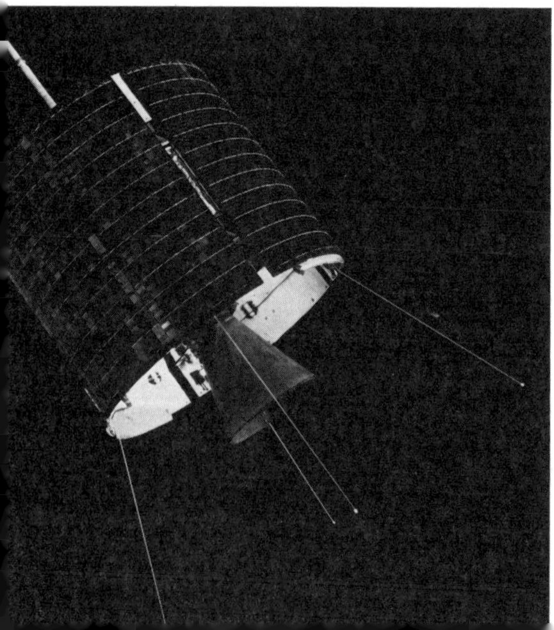

Early Bird on the outside.

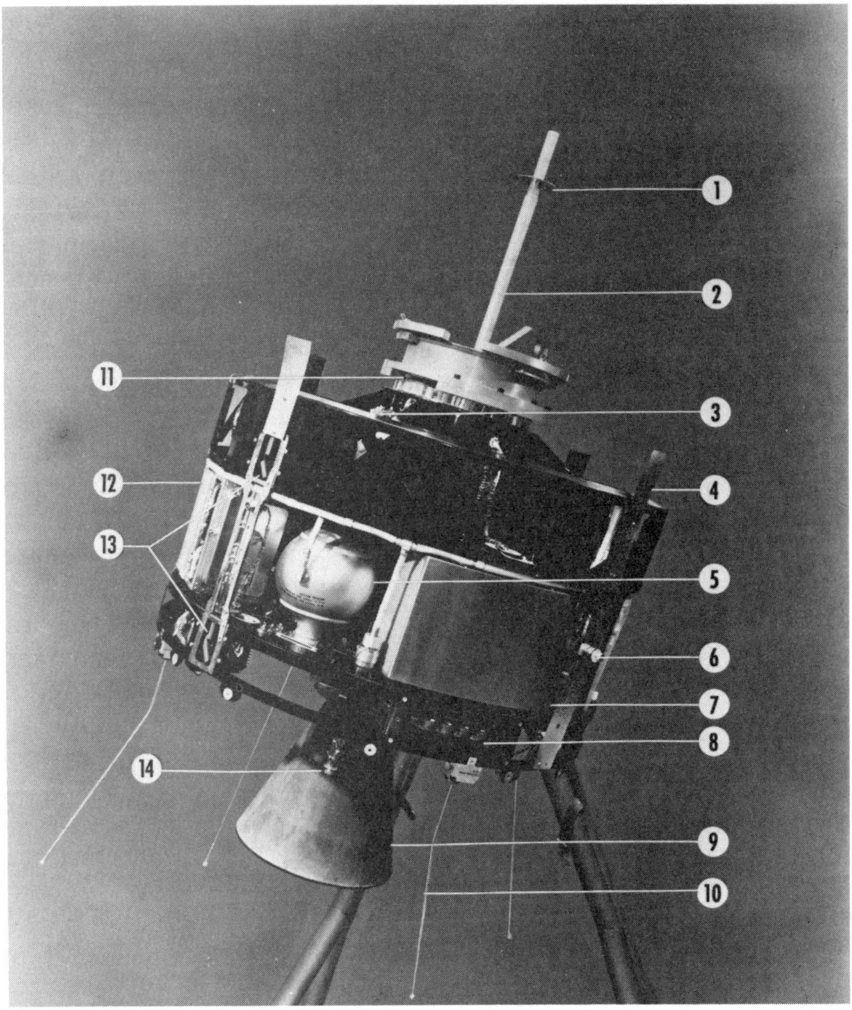

Early Bird on the inside . . .

1. *receiving antenna reflector*
2. *communications antenna*
3. *traveling wave tube*
4. *thermal shields*
5. *electronics*
6. *radial peroxide jet*
7. *transponder-receiver*
8. *nickel-cadmium batteries*
9. *apogee motor nozzle*
10. *telemetry antennas*
11. *separation interface for Delta rocket*
12. *encoder-decoder*
13. *sun-sensors*
14. *axial peroxide jet*

ing its normal complement of electronic and photographic gear, it would probably have been disabled.

Late in 1981, reports reached the West that the Soviet Union

had put an antisatellite killer station into low orbit. The most sophisticated antisatellite weapon in the world today, the new battle station is equipped with clusters of podded miniature attack vehicles, which are outfitted with the infrared homing systems that the Russians have long been perfecting. The presence in space of this station, capable of destroying multiple enemy spacecraft, will make detection of future antisatellite tests more difficult. Hitherto, U.S. early warning satellites and radars could pick up on ground-based booster launches, thus alerting surveillance personnel to any antisatellite test in progress. Given the new Soviet air-to-air launching capability, the United States will have to rely on space-to-space reconnaissance or ground-based radar to detect antisatellite tests.

Like the United States, the Soviets are actively developing space-based laser and particle-beam weapon technology. The Russians are reportedly well ahead of the United States in both areas. (For Soviet advances in particle beam weapons see chapter 6.)

U.S. ORBITAL WEAPONS

In 1978 the Carter administration, concerned by the "troublesome" gap in U.S. orbital weaponry development, announced an intensive effort in that domain. The most promising present defense against Soviet hunter systems consists of what experts describe as a "high-technology interceptor using a miniature vehicle that directly impacts the satellites." Manufactured by Vought Corporation, the device would be carried above 100,000 feet by an F-15 jet fighter, then launched to home in on its target. DOD is also working on ways to harden U.S. spacecraft to survive antisatellite aggression. In October 1981 the Soviets, in a petition outlawing all space-based weapons, cited the Vought missile as a violator of such a potential ban.

SATELLITE WARS

The importance of the satellite-bearing shuttle and the space-based killer battle station in U.S. and USSR defense programs attests to the changing role of satellites today. Since Sputnik inaugurated the space age in October 1957, over 3,000 satellites have been launched. Of the approximately 300 presently in use, 200 serve a primarily military function. At this writing **more money and research hours are being dedicated to the military application of satellites than to all civilian space functions combined.**

Satellites are employed militarily for:

* electronic data collecting.
* C^3I relay linkage.
* detection of nuclear explosions.
* conveyance of almost three fourths of all U.S. diplomatic and military communications, including talks on the Moscow-Washington hotline.
* global spying through a variety of hi-tech sensors.

The top-secret National Reconnaissance Office, which coordinates clandestine satellite survey missions with the help of its $2 billion annual budget, now uses satellites routinely to gather data, as do the CIA and other intelligence agencies. C^3I experts also back satellite use, believing that space-basing would place centralized commands in regions of comparative safety

NASA artist's rendering of an unmanned "laser battle station." Some defense experts believe that with only four laser battle stations in space the USSR could shoot down an entire fleet of high-altitude bombers and most tankers.

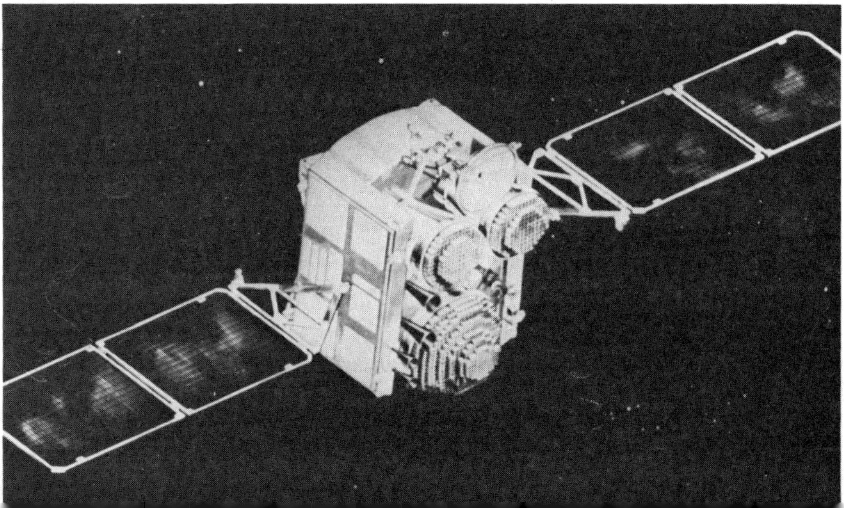

during global conflict. Strategists cite the use of advanced satellite reconnaissance to facilitate such feats as the demolition of targets deep within the USSR by U.S. submarine-launched missiles. At the same time, some analysts fear that parties aware that the precise location of their nuclear weapons was known to the enemy might feel constrained to use those devices before it was too late—in other words, *satellites could increase the possibility of relatively minor conflicts escalating into nuclear war.*

Whatever the pros and cons of space-basing, it is too late now to turn back. As Pierre Kohler puts it, "We are involved in more than an arms race. We are actually part of a relentless war between two superpowers to obtain supremacy in space—an ideal control platform from which to direct war on earth."

From a top U.S. weapons scientist, a Proposal:

"We should move with urgency to conclude a bilateral treaty (and then supplement it with an international accord) to ban destructive attacks on satellites, and to forbid the emplacement of instruments in space capable of projecting damaging levels of radiation."

Richard Garwin of
Harvard University and
IBM's Thomas J. Watson Research Center

Proposal made at the annual meeting of
American Association for the Advancement of Science, January 4, 1981

WAR IN SPACE:
Two Views

"The Trojan Horse was heroic in its time. The so-called killer satellite could be similarly heroic, with still more profound influence on world history. We will need major, simultaneous scientific advances to deploy such a system. But before that, we would need acceptance at the top that war can no longer be started, fought and won without considering space first. The huge military assets already in orbit on both sides hold the keys to winning—or losing—any future war.

"A fleet of orbiting, 5–6 megawatt laser 'battle stations' could sanitize space in 5 or 6 hours. DARPA [Defense Advanced Research Project Agency] is working on these power levels now. Such a device would neutralize satellites, and would also have look-down/shoot-down to deny tactical, AWACS-type battlefield control or bomber/cruiser-penetrations, since all of them would be forced down to, say 10,000 feet from their 35,000 feet working altitudes.

"Cost? I estimate $2 billion or more to orbit a demonstrator, and $400–500 billion to deploy a credible, 12-satellite system. Twenty or thirty years is a reasonable time frame. And yes, the Soviets confront the identical physics."

Ray Capiaux
Vice-President Research and Development
Lockheed
Palo Alto

"Rather than welcoming the opportunity to turn space into a boxing ring for a series of fights for which we cannot afford the admission, we should use our human and material resources more effectively to maintain our security and well-being. One might suggest that we let the other side waste its efforts on weapons in space while we counter with feasible, comparatively low-technology rockets, but experience shows that technological optimism, contractor pressure, and fear of asymmetry can result in costly and potentially destabilizing contests, to the detriment of our real military needs."

Richard Garwin of
Harvard University and IBM's Thomas J. Watson Research Center

The Sprint, an antimissile missile, launched from an underground launching cell.

chapter 8

Missiles: *The Precision-Guided Weapons*

Shortly after the termination of the 1967 Middle East war, a series of fat-bodied, finned missiles skimmed 10 miles over the Mediterranean to slam into the Israeli destroyer *Elath.* The delta-winged projectiles, launched from a 100-ton carrier, were identified as Russian-designed Styx, 20-foot rocket-propelled devices with 1,000-pound high-explosive warheads and radar-based terminal-guidance systems. **The destruction of the *Elath* marked the first sinking of a ship by unmanned homing missiles.**

On April 23, 1972, South Vietnamese tank crews tried to evade the erratic flight patterns of Russian-made Saggar antitank missiles. But by midmorning, according to a U.S. report, "all officers of the 3rd Squadron had been killed or wounded and three M48A3 tanks had been destroyed." The next year, during the Arab-Israeli war, the Saggar struck again when two-man Egyptian infantry teams opened containers resembling small suitcases and launched the 25-pound rockets. **With chilling dispatch the missiles' high-explosive warheads reduced Israeli battle tanks to charred and twisted skeletons.**

Both the Styx and the Saggar belong to a group known as "smart" weapons: conventional (nonnuclear) munitions that can be remotely guided to target or be homed in automatically. Fired from land, air, or sea, this remarkable type of munition is able to find and isolate targets with near-human canniness, track them with tenacious agility, and destroy them at a blow. It is no wonder that the Pentagon views the forthcoming generation of U.S. precision-guided weapons as one of the stars of its future arsenal. William J. Perry, former Under Secretary of Defense for Research and Engineering, says the new missiles will "revolutionize warfare." They are expected to change combat scenarios in two prime areas—tank warfare and fighter aircraft activity. In tank warfare the kill power of advanced antitank missiles would help NATO forces counterbalance Warsaw Pact tank supremacy in Central Europe (see chapter 4), while reducing U.S. dependence on expensive battle tanks. (The M-1 main battle tank is four hundred times more costly than the TOW [tube-launched, optically tracked, wire-guided] antitank missile and the tank/antitank-missile price discrepancy will increase as missiles carry and release dozens of mini-munitions, each capable of homing independently to target.) As for fighter aircraft, smart missiles

The Titan I, from the time the launch controller initiates the launch sequence to the instant just prior to lift-off from Vandenberg Air Force Base, California. The missile is housed in a concrete and steel underground silo hardened against nuclear attack. When launch is ordered, the 200-ton silo doors open, the missile is raised to the surface by an elevator and is launched within a matter of minutes.

could relieve them of such functions as dogfighting and low-level penetration sorties. Former Under Secretary of the Army Norman Augustine predicts: "**By the end of the decade the computers in missiles will come very close to comparing with the human brain. Our missiles will be not just smart but brilliant.**"

A HISTORY OF THE DEVELOPMENT OF GUIDED MISSILES

At the time of World War I, Orville Wright, E. A. Sperry, and Charles Kettering designed and tested the first U.S. guided missile, a miniature aircraft to be used to bomb targets. Although the diminutive plane never saw combat, it established the importance of radio control as a means to remotely control missiles during flight. Research in radio-controlled aircraft continued throughout the 1920s, and by December 1941 American models were sufficiently developed to be considered for military use. In fact, television-sighted, explosive-filled B-17 bombers unfit for further service were crash-landed by remote control over Germany in 1944. (Both the United States and Germany experimented with television-guided munitions well before the scheduling of regular commercial television emissions.)

Other U.S. guided munitions developed in World War II included GBs (guided bombs) and VBs (vertical bombs). Although the 109 one-ton winged GB-Is used over Cologne were in fact unguided, later models became increasingly sophisticated, some using radar technology. VB munitions, in contrast, were guided from the start. Early models of the free-fall, wingless missiles were equipped with tail guidance. The VB-I, the sole VB to be used extensively in the war, was called the Azon since it could be guided by azimuth only via a radio receiver in its tail. The radio-guided VB-3, or Razon, had both range and azimuth steering but saw no wartime action. Later VB models were equipped with TV control and radar homing. The last in the series, the VB-13, or Tarzon, a mammoth 6-ton, 21-feet-long missile, was used in Korea.

The first American homing missile, a glide-supported bomb called the Bat, was employed against the Japanese navy in the Pacific. The sophisticated winged missile contained pulse radar in its nose (see chapter 11) and homed in on reflections from target ships.

Also worthy of note, historically, is the British surface-to-air missile Brakemine, which utilized the beam-rider concept. As early as 1925 there had been talk of guiding a missile along a searchlight beam by means of light-sensitive cells in its tail fins. The more sophisticated Brakemine was intended to ride a radar beam locked onto a target. Undertaken during the 1940s, the Brakemine program was eventually scuttled due to lack of funding, although the missile was gradually developing into a usable weapon.

The Germans were extremely active in missile development during the 1930s and 1940s. By 1936 they had decided to make guided missile research a major priority, spending $40 million on the so-called Peenemünde Project, a rocket laboratory where specialists from the German Rocket Society labored throughout the war. The two German missiles that saw wartime service were self-guided: the V-1, or buzz bomb, a pioneer cruise missile; and the revolutionary surface-to-surface V-2, the first missile to exceed the speed of sound. *The 46-feet-long V-2,*

Construction of the Minuteman missile launch site, at Whitman Air Force Base, Missouri, in 1962.

against which there was no defense, soared to 60 miles, the greatest height yet achieved at that time by a man-made object.

By war's end, the Germans had close to 150 guided missile programs in operation. The following attest to the breadth of German technology:

> *The X-4* established the air-to-air missile as a viable concept and also proved the efficacy of wire guidance, a system later used widely for antitank weapons. By late 1944 about 13,000 of the 6.5-feet-long, swept-wing missiles (controlled by electrical signals sent along the fine wires that wound down from the wings) had been produced and many hundreds had been tested. None is known to have reached combat units.
>
> *Rheintochter* (Rhine Daughter), a large and ambitious surface-to-air missile had 4 braced swept fins and 6 fixed wings. Rheintochter measured some 20 feet long and over 16 feet in diameter. Although never used in battle, it provided flight-proven guidance hardware for future generations. Rheintochter used radio CLOS (Command to Line-Of-Sight) guidance, a system used frequently in early German army and Luftwaffe missiles. CLOS operated by tracking enemy aircraft with steerable telescopes, firing the missiles and keeping them on target through steering commands sent by radio link. Flares on opposite wings helped identify the Rheintochter.
>
> *Wasserfall,* a 25-feet-long supersonic rocket designed to intercept aircraft, employed radar technology. Radio guidance was executed by a ground-based operator who watched sight lines on a complex display in which target and missile were tracked by pencil beam radar. The operator's job was to steer the missile to keep the sight lines coincident, then press a button to detonate the 200-pound warhead. Plagued by design changes and accidents, the Wasserfall program was abandoned in 1945 after several dozen full-size models had been fired.

The Japanese, although far behind the Germans technologically, added two distinctive if macabre guided bombs to the

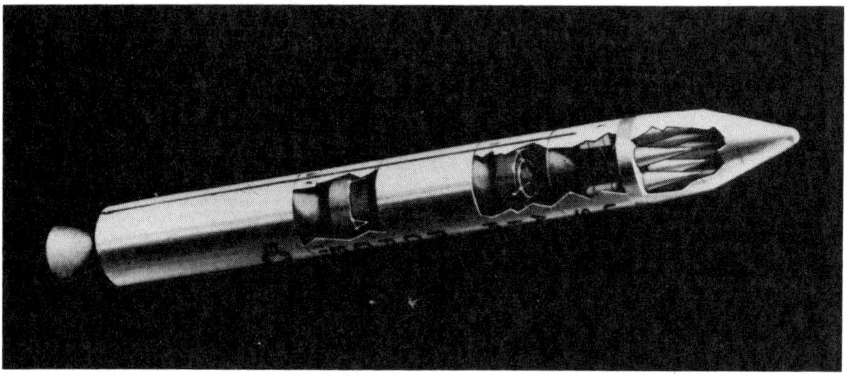

The MX missile is about 70 feet long, 92 inches in diameter and weighs 192,000 pounds.

World War II theater. The first, the kamikaze (heavenly wind) or suicide plane, was crash-dived by its pilot onto enemy decks. The second, the *oka* (cherry blossom) bomb, dubbed *baka* (screwball) by U.S. forces, was a rocket-propelled manned glider carrying over a ton of explosives. It was released from the underbelly of bombers to home in on ships at speeds of up to 600 miles per hour.

Studies show that the early guided missiles were far from precise. A U.S. postwar survey reports that only about 20 percent of the bombs aimed at precision targets during World War II "fell within the target area." (The radius of a "precision" target at that time was defined as extending 1,000 feet from the aiming point.)

FIRE AND FOLLOW: THE "EDUCATED" ANTITANK WEAPONS

The first truly precision-guided weapons, antitank devices, appeared in the mid-1950s following developments in guidance technology and small rocket motor design. Using the German X-4 as a model, the French created the wire-guided SS10 and SS11, which were succeeded by the British Vigilant and the German Cobra. In these first generation systems, operators tracked both missile and target and manually guided the projectile to its destination by means of a joystick.

A major breakthrough in smart weaponry occurred in the early 1960s when the French patented TCA, a semi-automatic guidance system that eliminated the need for a joystick. Operators could now guide missiles home by centering their targets in the crosshairs of an optical sight. An infrared sensor mounted alongside the sight automatically tracked an infrared source at the rear of the missile. Any deviation from target generated a command signal in a computer that relayed it along the wires playing out from the missile to the missile control system. The American TOW antitank missile is an example of this technology. *To date 275,000 TOWs have been sold to 33 countries—an all-time record in guided missile production.*

FIRE AND FORGET: SMART WEAPONS COME OF AGE

The invention of automatic homing, advances in imaging infrared and sensor technology, and the capacity of microcomputers to translate signals into steering instructions with ever-increasing speed have contributed to the birth of a new generation of precision-guided weapons. Operators no longer need to follow missiles to target optically. There are also alternatives to the trailing guidance wires that give antitank missiles added vulnerability. Once operators have trained their sights on target, any de-

Designed for intercepting intermediate and intercontinental ballistic missiles, the 27-foot Sprint goes from launch to intercept in bare seconds.

viation from the sight line (a laser beam, for example) is measured by an apparatus in the missile that generates the proper guidance commands and applies them to the missile guidance system. Moreover, the time is in sight when fighter planes, helicopters, and other firing platforms will no longer need to help designate targets with radar and laser light, exercises accomplished at considerable risk. Missiles will be fired, then forgotten, by sources that can "shoot then scoot" to safety.

The Maverick One of the most widely deployed of the emerging family of fire and forget weapons is an air-to-surface munition said to have scored direct hits in almost 90 percent of its test firings. Versions of the missile (which became fully operational in the 1970s) have been developed using three different guidance systems: television, infrared, and electro-optical/laser. In the first model a tiny TV camera in the missile's nose relays a picture to a television screen aboard the attack plane. After selecting a target, the pilot moves a set of cross hairs to its position on the screen. This locks the missile onto the target, to which the munition will now proceed automatically. The pilot is free to attack a new target or to move out of enemy range. TV-guided Mavericks have a disadvantage compared to other models because bad weather and darkness can degrade their performance.

Infrared-guided Mavericks utilize sensors to detect variations in the amount of heat radiated by various objects, differentiating between live objects and "hulks." Electro-optical systems guide the third type of Maverick toward laser light that has been beamed onto targets from ground-based or air-based sources.

Close-up of short-range attack missiles (SRAMs) mounted under the wing of a B-52.

Other landmark advanced precision-guided weapons include:

Copperhead and **Hellfire** The Copperhead, an antitank device, is launched from a howitzer. During testing it has proved its lethality at ranges up to 10 miles. The air-to-surface Hellfire (Heliborne-Launched Fire and Forget) has been called "the U.S. Army's next generation anti-armored weapon." (For more about these sophisticated laser-guided missiles see chapter 5.)

Sparrow This powerful and maneuverable radar-guided air-to-air missile has spawned an entire family of weapons including the air-to-surface *Shrike* and the ship-launched surface-to-air *Sea Sparrow.* The latest Sparrow, the AIM-7E, travels at more than 2,300 miles per hour. An all-weather device, it uses movable wings to adjust its flight to climb or dive.

Sidewinder Equipped with infrared homing, the air-to-air Sidewinder is noted for its simplicity. Developed in the 1950s, it has fewer than 24 moving parts and fewer electronic components than the average radio. New models include radar homing.

HAWK (Homing-All-the-Way Killer) A surface-to-air missile, Hawk seems to have merited its name with a kill-per-engagement rate of 96 percent in combat in Southeast Asia and the Middle East and tests at home. Hawk is equipped with multiple radars and highly sophisticated command and control and data processing systems. After isolating moving targets from ground clutter, Hawk computes their bearing speed and range and can engage them at latitudes as low as 100 feet.

Phoenix The most sophisticated air-to-air missile in the world, the Phoenix has a 100-mile range. When it arrives within 10 miles of target, it activates its own terminal homing radar. A fire control radar aboard the launching aircraft can track 20 targets while guiding 6 Phoenix missiles to separate destinations.

Firing the Redeye missile.

SMART BOMBS

The so-called smart bombs are a further example of emerging fire and forget technology. This family of automatically homing munitions derives from the Paveway program instituted by the U.S. Air Force in 1966. (For more about the laser-guided Paveways see chapter 5.)

A second laser-directed munition, the lethal Rockeye antitank cluster bomb, produces fragments that cut through armor at speeds of up to 4,000 feet per second.

The Walleye, a television-guided smart bomb, was developed concurrently with the Paveway. In 1969 the Navy called it the most accurate and effective air-to-surface conventional weapon ever developed. Present versions have optional imaging infrared sensors and high-explosive warheads weighing more than 1,000 pounds.

The HOBO (Home Bomb System Munition), another important smart bomb, is electro-optically locked onto target by the pilot. It then homes itself in. Like the Paveway, the HOBO consists of a series of add-on modules fastened to standard "iron bombs." HOBOs were used in Southeast Asia in 1969, where they demonstrated good accuracy. Later models have increased numbers of bomb payloads and greater glide capacity. HOBOs are expected to make direct hits each time with minimal exposure to enemy defenses.

How effective are these new munitions?

Smart bombs equipped with advanced laser- and television-guided devices are purported to have destroyed 106 bridges in three months during 1972 in North Vietnam. Several of the installations thus demolished had withstood multiple previous attacks during which smart weapons were not used.

COUNTERMEASURES TO PRECISION-GUIDED WEAPONS

Despite their advanced technology the performance of today's smart weapons systems can be hampered by:

Weather Fog, snow, and rain can limit the effectiveness of television guidance systems. Hence infrared and laser guidance may be preferred in poor weather, although laser light is affected by clouds.

Outmaneuvering Some wire-guided missiles take 15 times longer to reach target than a high-speed tank round re-

The Sparrow

quires to strike home. Hence, tank crews trained to approximate the whereabouts of missile installations can strike at munitions crews first. But as the new antitank missiles gradually outstrip the tank gun in range, tanks will be forced to rely more heavily on long-range artillery and other supporting forces for defense.

Camouflage Tanks can be hidden behind smoke spewed from smoke grenade launchers, and laser and infrared homing devices can be confused by misleading flares and lights.

Improved armor Better antitank warhead penetration has led to the creation of stronger armor. (Some warheads can penetrate steel armor to a depth of five times the shell's diameter.) Two leading new armors are the British Chobham, which sandwiches ceramic material and fabric between layers of aluminum and laminated steel in order to dissipate kinetic energy, and the French appliqué, which has removable and addable plates.

Vulnerability of spotters While beaming laser light onto targets, aircraft and artillery forward observers are dangerously exposed to enemy attack.

As smart weapons grow increasingly accurate, mobile, versatile, and able to penetrate armor, they are expected to overcome present obstacles. Thanks to advances in semiconductor technology (viz. the silicon-chip brains of "brilliant" weapons), future guidance systems will exceed the signal processing and sensory capabilities of present models several thousand times over. By 1985 the armed services will have spent over $80 million on the development of VHSIC (very high speed integrated silicon chip circuitry). By 1990 VHSIC technology will make it possible for such weapons as air-to-air missiles to execute right-angle turns as they streak after their prey.

IN THE WORKS: THE NEW BRILLIANT WEAPONS

According to Norman Augustine, a vice-president of Martin Marietta Aerospace who used to work at the Pentagon, missiles now being developed will be able to make lists for their signal-processing computers to evaluate. "I see a tank, a bridge, and an armored personnel carrier," they might report. A missile computer programmed for antitank activity, for example, would select the tank as the first priority target.

A look at precision-guided weapons currently under development gives credence to predictions that future missiles will need humans only for launch and to inform them of an enemy presence "somewhere out there." The missiles will do the rest.

AMRAAM (advanced medium range air-to-air missile) Also called "the beyond visual range" missile, AMRAAM is expected to replace the Sparrow by the mid-1980s. This joint Air Force–Navy program will produce a weapon that is smaller, lighter, and cheaper and that offers a greater kill probability than its predecessor. The high-speed fire and forget weapon will acquire targets at extreme range without the need for constant target illumination by the launching aircraft. An active radar guidance system will bring the missile home, beaming in on the reflection of radiation the missile has emitted toward the target. AMRAAM will weigh 300 pounds, as compared to the 540-pound Sparrow, and be narrower and shorter. The government has supplied AMRAAM contractors with data bases whose information will make it possible for warheads to attack multiple targets. Following missiles will be programmed to skip over targets already engaged by lead missiles and will be capable of isolating targets of their own.

Assault Breaker This highly ambitious U.S. Army program is designed to wipe out tanks deep within enemy lines. A new, high-power, all-weather, airborne radar called Pace Mover will detect and track mobile targets and relay their positions to a missile battery which will then launch rockets. During flight, the projectiles' course will be adjusted by radar command guidance. Once over target, the warheads will release a cluster of smart submunitions whose infrared or millimeter wave homing sensors will acquire and lock onto enemy tank turrets.

Hypervelocity missile Hypervelocity technology has been likened to harnessing a tornado to drive a straw through a building or tree. **Hypervelocities are small, lightweight, air-, sea- , and land-launched missiles that will achieve velocities of up to 5,000 feet per second.** Their lethality would derive from the kinetic energy they develop during flight. At anticipated speeds they would defy detection. Engineers at the Vought Corporation, which is pioneering this new technology, say that missiles 3 feet long and about 2 inches in diameter have been fired through armor plate, creating holes the size of grapefruits. Models now under study contain tiny laser/radar guidance packages in their noses.

MLRS (multiple launched rocket system) This antitank device will fire rockets containing 600 submunitions to a distance of 20 miles. Twelve missiles can be fired per volley, saturating an area the size of six football fields with armor-piercing fragments. The MLRS was scheduled for deployment in 1982. The Army is slated to purchase some 360,000 rockets by the mid-1980s.

SADARM (sense and destroy armor) Scheduled for deployment in the mid-1980s, this antitank system will contain passive microwave sensors judged "effective against all known countermeasures" by their manufacturers. (Passive sensors pick up signals originating in the target.) Munitions launched from howitzers will travel 20 miles or more before releasing three submunitions. Each submunition will parachute toward earth, relying on its tiny terminal guidance systems to direct it toward ground-based microwave-emitting targets.

STAFF (smart target-activated fire and forget) A projected antitank missile, STAFF will fly over tanks within a 2-mile range at an altitude of 100 feet or less. Fired from the shoulder or from a vehicle, projectiles will detonate upon sensing a target below, riddling tank turrets with armor-piercing fragments.

SWERVE (Sandia winged energetic reentry vehicles) This hypersonic tactical missile system is still in the experimental stage at the Sandia Laboratories. Missiles will be surface-launched into the exoatmosphere at speeds of several thousand miles per hour. Upon reentry they will strike targets within a 1,000-mile radius. Experts believe that SWERVE will provide a fast-reaction defense system for use primarily against cruise missiles, ships, aircraft, and fixed or mobile ground targets. The nature of the sensor guidance package has not yet been defined.

WAAM (wide area antiarmor munition) Sponsored by the Air Force, this air-to-surface program comprises three smart weapons. Elements include:

ACM (antiarmor cluster munition) Initial production of this small weapon, the most advanced segment of the WAAM program, is projected for 1983. ACMs are released in clusters that, like SADARM systems, parachute to earth. Upon contact with tank tops or the ground, the munitions explode, sending one slug downward and two to the sides.

ERAM (extended range antiarmor munition) This missile is designed to be dropped atop or in the path of armored targets. Highly developed built-in signal processors and sensors allow the missile to discriminate between armored tracked targets and other kinds of vehicles so that it can limit attacks to the former.

WASP A 100-pound minimissile to be launched from an aircraft pod, Wasp will automatically acquire and home in on targets. It will have two types of guidance: microwave seeker, which identifies radiation emitted from the target; and infrared sensor, which discriminates among heat sources. The infrared system will be programmed to lock onto the shape of a tank. Like AMRAAM, Wasp will bypass a primary target already engaged by a preceding missile and lock onto the next priority target. Its speed and low flight path will make it almost impossible to shoot down.

The controversial B-1, a bomber that can launch up to 24 cruise missiles at distant targets.

chapter 9

The Nuclear Triad: *Our Three-Legged Defense Strategy*

For the last two decades the United States has maintained a three-part strategic weapons system dedicated to delivering enough destructive force to paralyze any enemy. The air-based leg of this nuclear triad consists of long-range strategic bombers; the land-based segment comprises intercontinental ballistic missiles (ICBMs) fitted with nuclear warheads; and the sea-based branch is composed of submarine-launched ballistic missiles (SLBMs). Although any one of these systems is capable of annihilating the earth as we know it, each is believed to be vulnerable. For this reason the triad concept remains a cornerstone of our policy of deterrence. The way DOD figures it, should U.S. forces be caught in a normal alert situation after a massive Soviet first strike, each leg of the triad would be able to wipe out at least 75 percent of the Soviet industrial base, two legs in tandem could eradicate 65 percent of such targets plus all military targets (excluding Russian ICBM silos), and the three legs together could destroy more than four-fifths of governmental centers as well as industrial and general-purpose force targets, *while keeping 1,000 weapons in reserve.* In spite of such readiness the early 1980s are being marked by a massive U.S. nuclear buildup.

THE AIR LEG

The B-52 Boeing Stratofortress

A long-range manned penetrating bomber, the B-52 is the United States' oldest strategic weapon. In the late 1950s and early 1960s, 700 of the aircraft were produced. Close to 350 modified improved B-52s are currently in service, equipped with advanced electronic countermeasures. Measuring 160 feet in length, with a 185-feet wingspan, and carrying a crew of six, the B-52 attains a top speed of 650 miles per hour. About 30 percent of these bombers are on alert in peacetime. Most models carry a mixture of short-range attack missiles (SRAMs) capable of delivering a 200-kiloton warhead to a distance of over 100 miles at triple the speed of sound, and up to four gravity bombs, with yields of from 1 megaton to 200 kilotons.

In spite of their astonishing resiliency the B-52s are likely to be phased out in the 1990s due to age and a decreasing ability to permeate advanced Soviet radar, surface-to-air (SAM) missiles, and fighters. The problem of bomber penetration, however, has been greatly alleviated by the fitting of 170 B-52s with 20 cruise missiles each: 12 cruise missiles will be carried under the aircraft's wings and 8 in a rotary launcher in its weapons bay. (For more on cruise missiles see page 123.) By 1987 there will be close to 3,500 of these strategic nuclear projectiles aloft, with ranges of 750 to 2,000 miles. The B-52 will thus serve largely as a remote launch platform.

It is believed that 40 B-52s could deliver enough cruise missiles to wipe out half of the Soviets' military target base. Fewer than 60 could knock out 80 percent of USSR industrial targets.

The B-1

The B-1 was the most hotly debated weapon in the Reagan administration's proposed 1981 arms program. The four-engine swept-wing supersonic bomber, carrying a crew of up to five, was conceived to replace the venerable B-52. One of the aircraft's main attractions is its alleged ability to penetrate Soviet air defense by flying as low as 200 feet off the ground. This capacity, along with its state-of-the-art conformation, gives the B-1 a radar cross section about one tenth as "visible" to radar as that of the B-52. *Its advanced electronic equipment is said to include a device that can generate false images of the bomber on a gunner's radar, thereby creating confusing multiple targets.*

The B-1 has been surrounded by controversy since the first model, built by Rockwell International, took to the air in 1974. Throughout the 1970s its fate hung in the balance while successive administrations and Congress debated the necessity for manned bombers in an age of long-range nuclear missiles. Although production of the B-1 was ordered by the Ford administration during its last weeks, President Carter canceled the program six months later, after three additional B-1 models slated for test flight had been constructed.

The B-1 can drop nuclear or conventional bombs as it skims over enemy territory, or, like the B-52, it can stand off and launch up to 24 cruise missiles at distant targets. Fleeter, lighter, and smaller than the B-52 (147 feet long with a 137-feet wing spread extended and 86-feet span swept back), it can carry five times the B-52's payload.

As 1981 ended, 100 B-1s were scheduled for production, the first models to be ready by 1986. Pro B-1 factions stress that the plane's radar-elusive qualities have already been proven in testing, that the United States needs a manned bomber, recallable in case of error, and that, as Senator Gary Hart of Colorado puts it, our Air Force will benefit from an aircraft that will create a new "jujitsu puzzle" for the Soviets to contend with.

GLCMs

The controversial Tomahawk ground-launched cruise missile has a nuclear warhead with a yield in the "upper kilotons."

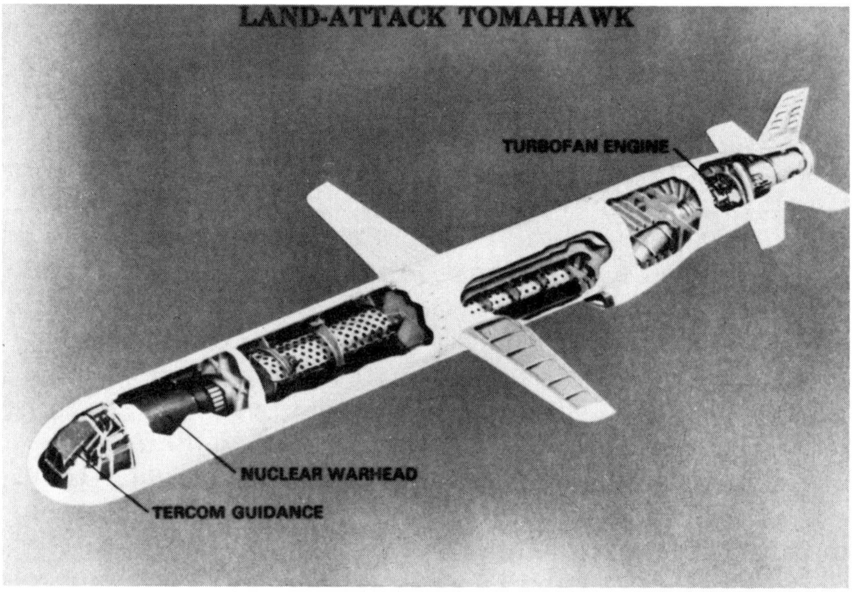

Artist's cutaway of the Tomahawk.

NATO has requested that 464 of these medium-range weapons and 108 Pershing II missiles be stationed in five Western European countries by 1983. According to Richard Perle, Assistant Secretary of Defense for International Security Policy, such ground-basing of U.S. missiles abroad will serve as "an expression of alliance solidarity" and reflect a "willingness to share the risks and burdens of providing for a defense against Soviet nuclear weapons."

In November 1981 President Reagan announced what has come to be known as the "zero option" plan, stating that the United States was prepared to cancel projects for the 572 European-based missiles if the Russians would dismantle comparable arms. These include Soviet SS-4s, SS-5s, and particularly the formidable SS-20s, deadly surface-to-surface missiles with 3,000-mile ranges and three warheads, each boasting an estimated 150-kiloton yield. United States–USSR dialogues concerning GLCM and related arms limitations promise to be complex and protracted.

SLCMs

According to Admiral Harry Train, commander of the Atlantic Fleet, ship- and submarine-launched SLCMs (Submarine-Launched Cruise Missile) will initiate "the greatest change in the role of navies that has ever occurred," serving as the "technological basis for a revolution in the concept of sea power."

Submarine-launched Tomahawks with conventional warheads will be able to travel 700 miles over land and 250 miles over water at speeds of 500 miles per hour.

Presently being developed is a Tomahawk Cruise Missile Capsule Launcher Subsystem with a 1,500-mile range. Other sea-based options include the launching of missiles from cruisers, reactivated battleships and carrier-based attack planes that could lob missiles at land-based targets from a distance of 700 miles.

The Cruise

The introduction of cruise missiles into our national arsenal has sparked opposition in some quarters. In an "Op Ed" piece in *The New York Times* (December 1981), Alan B. Sherr, president of the Lawyers Alliance for Nuclear Arms Control, suggested that Reagan's decision to deploy the cruise is likely to have a greater effect on our national security than deployment of either the B-1 bomber or the MX missile (see page 126).

Of major concern is the fact that the missile's small size and mobility make it easy to hide from satellite and other types of surveillance. Although the United States might not resort to disguising its missile strength, suspicious nations could both refuse to negotiate the reduction of a type of missile whose presence could not be measured, and step up their own production of that arm in order to forestall a weapon gap. Also worrisome is the cruise missile's relatively low cost (approximately half a million dollars per missile), which will make likely the increased proliferation of this weapon. Sherr cautions that the United States must not repeat the mistake it made in the 1970s, when it rejected the chance to outlaw multiple independently targeted reentry vehicles (MIRVs), deciding instead to capitalize on the fact that we had the vehicles and the Russians did not. The very fact that we have developed and tested the cruise well in advance of the Soviets, says Sherr, gives us the leeway to ban while we can.

Cruise missiles are self-guided. They estimate their bearings through a technique known as TERCOM (TERrain-COntour-Matching). TERCOM compares ground-based data picked up by the missile's sensors with on-board computerized altitude maps made via satellite and stored on magnetic tapes. It then alerts the inertial guidance system to make any adjustments necessary to keep the missile on course.

Stealth

In 1980, then Secretary of Defense Harold Brown fired the public imagination by announcing a multibillion dollar research and development program aimed at producing a penetrating manned strategic bomber that would be "virtually invisible to radar." Little information about the highly classified aircraft is available to the public. Indeed, the former chief of Air Force Intelligence, George Keegan, warned in October 1981 that "there is no such thing as a Stealth bomber. . . . It's a piece of paper, it's design work." Although many sources maintain that the advanced-technology aircraft will be available for deployment by the early 1990s, that timetable could be premature.

While the idea behind Stealth technology is relatively straightforward, the procedures necessary for pulling off radar invisibility are not easy to execute. For a plane to be rendered "invisible," discontinuities on the aircraft skin, such as holes or protuberances, must be eliminated in order to reduce the energy that is reflected back as tell-tale signals. Immediately design problems will develop, such as figuring out where to place the bomber's jet inlets.

Other radar eluding devices might include electronic countermeasures that drown out or confound hostile radar emissions, minimization of the aircraft's dimensions, special coatings that partially absorb radar wave energy, lessening the reflection back to the transmitting source and a recently perfected sophisticated technique called destructive interference, which eliminates radar returns by reflecting radar waves partly from the aircraft's outer coating and partly from material beneath the bomber's surface in a way that causes the two sets of waves to cancel each other out.

THE GROUND LEG

Titans and Minutemen

The second leg of the U.S. nuclear triad consists of 1,054 silo-based intercontinental ballistic missiles (ICBMs), 54 Titan IIs, 450 Minuteman IIs, and 550 Minuteman IIIs.

Our Titan IIs, operational since 1963, are currently siloed in the Midwest.

Their 10-megaton warheads make them by far the most powerful weapons in the U.S. nuclear arsenal. They can blast 1,000-feet-deep, 1-mile-wide trenches in the earth.

A Mark IV nose cone is mated to a Titan intercontinental ballistic missile in a silo at Vandenberg Air Force Base, California.

The Titan II's successor, Minuteman II, became active in 1965. Its dramatically smaller 2-megaton warhead owes its potency to greater accuracy. (Roughly speaking, a doubling in accuracy increases destructive power by 800 percent.) Both the Titan II and the Minuteman II have ranges of over 7,000 miles.

Minuteman III, based in the Dakotas and Wyoming, became active in 1970. Able to travel over 8,000 miles at speeds of up to 15,000 miles per hour, these missiles can reach Soviet targets within half an hour. Minuteman III was the first ICBM to be MIRVed. (See MIRVs and MARVs, page 128.) Its initial Mark 12 MIRVs contained three 170-kiloton-yielding warheads. The new Mark 12A delivers three warheads yielding 335 kilotons each. The Mark 12A is said to reduce the missile's theoretical margin of error when aimed at center target by well over a third—from 350 meters to less than 200 meters.

The MX

The centerpiece of President Reagan's strategic package, the MX—or Missile Experimental—has been in the works since 1965 when the Soviets deployed their first missiles believed to be capable of attacking and cutting off the capability of American ICBM launch-control centers. One of Gerald Ford's final budget actions was to recommend funds for full-scale MX engineering development. When it seemed apparent that the Soviets would have a new generation of ICBMs ready by the mid-1980s with the accuracy necessary to destroy U.S. silos, President Carter embraced the MX concept as a method of assuring that the United

A full-scale mock-up of the mobile MX intercontinental ballistic missile. The first MX flight test will be in 1983. Initial operational capability (10 missiles) is scheduled for 1986; full operational capability (200 missiles) is scheduled for 1989.

States would have "a secure strategic deterrent now and in the future."

The four-staged MX, the largest ICBM permissible under SALT, will be the most powerful missile ever developed by the United States: 71 feet long, 92 inches wide, and weighing in at close to 200,000 pounds at launch, it will deliver ten pre-aimed, electronically guided Mark 12A warheads, directed at different targets. Later models will carry warheads yielding up to 500 kilotons each.

Although some see the MX as the "hidden time bomb of the nuclear arms race," the considerable controversy over the missile has revolved mainly around where to put it, not whether to build it. Carter's plan to disguise the whereabouts of the missiles by shuttling them along 4,600 western-based shelters (called, originally, the "MX Racetrack") raised a furor with environmentalists and struck some politicians and defense experts as well as sheerly ludicrous. Carter's giant shell-game would have required up to 12,000 miles of new road construction, 100,000 billion gallons of water, much of it pumped from already receding desert aquifers, and the acquisition by the Air Force of some 5,400 land parcels, largely grazing land. In late 1981 Reagan decided to limit

Warheads for the MIRV missile (multiple independently targeted reentry vehicle).

MX deployment to 200 missiles and to place them in existing ICBM silos.

Proponents of the MX laud its accuracy, flexibility, and relative invulnerability as compared to the Minuteman III. Critics point to the fact that all missiles currently under development are tending to miss their targets due to "bias," a combination of uncontrollable factors including variations in the earth's gravitational pull and changing atmospheric effects at reentry time. Could this not, they ask, turn the MX's first-strike capability into what former Secretary of Defense Harold Brown calls "a cosmic roll of the dice"?

MIRVs and MARVs

MIRVing—the use of multiple independently targeted reentry vehicles—greatly augments missile lethality.

It allows the missile to shoot several missiles just as a six-shooter permits one man to shoot many.

MIRVed missiles carry two or more nuclear warheads in their final stage. Upon ejection each warhead goes for the target following its individually preprogrammed course. MIRVing is both efficient and effective. It also results in significant stretching of SALT-imposed arms restrictions. SALT I limited the number of ICBMs per nation but laid down no ground rules for how many independently targeted warheads each might carry. Even if SALT II, which would aim to limit strategic delivery systems throughout the triad, were to succeed in determining the number of ICBMs that can be MIRVed, there would be no restriction on the number of warheads per missile and no prohibition against upgrading their accuracy.

The United States pioneered MIRV technology, deploying the first MIRVed missiles in 1970. In 1973 the Soviets tested

MIRVs of their own and and initiated MIRV deployment two years later. The United States presently fits the Minuteman III with Mark 12 and Mark 12A warheads. The MX will employ the latter. Polaris, Poseidon, and Trident SLBMs (submarine-launched ballistic missiles) are also equipped with multiple reentry vehicles.

MARVs (maneuvering reentry vehicles) use an even more advanced technology. MARVed missiles can actively avoid hostile radar or weaponry by changing course in space. The MX may eventually be fitted with Mark 500 MARVs. The vehicles are presently plagued by limited accuracy due to their evasive maneuvering tactics, but future models may have terminal guidance systems that would allow them to strike within 100 meters of center target.

THE SEA LEG

The U.S. Navy's present fleet of over 40 SSBNs (ballistic missile submarines) serves as a mobile network of underwater pads from which to sea-launch long-range nuclear armed missiles (SLBMs).

When patrolling the deep, SSBNs and their missile payloads are virtually untargetable, making the sea-leg of the triad its most survivable arm.

At this writing the United States has 42 SSBNs, 41 Polaris and Poseidon submarines, plus the first of the new Trident-class subs, the *U.S.S. Ohio.* The Polaris and Poseidons were delivered between 1960 and 1967. The *Ohio,* the largest submarine outside of the Soviet Union, began sea trials in June 1981.

The Polaris

The Polaris, the first U.S. nuclear sub to be fitted with SLBMs, was launched in 1960, the same year that the Soviets put their first SLBM-bearing sub to sea. Ten of these 7,000-ton, long-range vessels, each fitted with 16 MIRVed Polaris A-3 missiles, remain seaborne today. The Polaris A-3 delivers three 200-kiloton warheads to a distance of 2,880 miles.

The Poseidon

The Poseidon SSBN, which succeeded the Polaris, currently has 31 ships afloat. The 435-feet-long submarine is 55 feet longer and 1,500 tons heavier than its predecessor, but its range (4,500 miles) and patrol days (70) are the same. Like the Polaris, it carries 16 missiles. The chief difference between the two subs lies in the Poseidon's more sophisticated missile, which became operational in 1971.

The Trident

Longer than the Washington Monument is tall, the 550-feet Trident submarine was conceived as a giant launching platform for silo-busting missiles. It is quieter than its predecessors, boasts improved communication and sonar systems, and can launch 24 missiles. Due to production mishaps the *U.S.S. Ohio* came in 30 months after deadline and $280 million over budget. Because of these costs and delays the schedules and specifications for future Tridents are not definite at the moment, although the Navy orig-

The U.S.S. Ohio, *at the naval base in Groton, Connecticut, was built by General Dynamics. The* Ohio *is the Navy's first nuclear-powered Trident missile submarine.*

inally planned to build 2 to 3 Tridents a year for an eventual total of 30.

The Trident I (or C-4) missile stemmed from a 1972 order by President Nixon to develop a "far more effective missile" than the C-3. Four months later the C-4 was in the works. Like its launching craft, the *U.S.S. Ohio,* the Trident I missile was plagued with monetary and technical mishaps, delaying the launch of the first Trident missiles until 1977.

Trident I has an impressive 4,600-mile range. Each of its MIRVed missiles will deliver 17 warheads. The proposed Trident II missile (D-5) will have an even greater range, traveling up to 6,000 miles. This will allow it to destroy Soviet ICBMs in their silos, something Trident I can't do.

As 1982 began, the Polaris fleet carried 480 warheads and the Poseidon fleet up to 6,368 more. By the mid-1980s, 6 Trident submarines, each bearing 240 warheads, may be able to add a whopping 1,440 nuclear warheads to the triad's missile-heavy sea leg.

"What's the difference if we have 100 nuclear submarines or 200. . . . You can sink everything on the oceans several times over with the number we have, and so can they. . . . Our leaders keep using scare words to get what they want."

Admiral Hyman G. Rickover,
father of the nuclear Navy,
in his official retirement speech, January 1981

A telescope used by the U.S. Army Satellite Communications Agency for tracking guided missiles.

chapter 10

Staging the War with Computers: *The World of C^3I*

A major cause for Defense Department concern is the distance between the continental United States and possible theaters of military operation. Southeast Asia, for example, is up to 12,000 sea miles away. While only a few World War II operations involved force projection of beyond 300 nautical miles, military planners today are required to project enormous force over vast distances, at tremendous speeds, and in the face of widespread political uncertainty.

A key element in assuring effective crisis management is adequate "connectivity"—quick, reliable communication between our National Command Authorities (the President, the Secretary of Defense, and their deputized alternates or successors) and the commanders of our strategic weapons units. In the late 1970s the U.S. Secretary of Defense decided to combine programs that manage command and control of our strategic and general-purpose forces with those involving computerized intelligence resources, placing all of these elements under a single management umbrella within the Pentagon. Thus, Command, Control, Communications, and Intelligence, or C^3I (pronounced "see cubed eye"), was born. According to the official Joint Chiefs of Staff definition, the functions of C^3I "are performed through an arrangement of personnel, equipment, communications, facilities, and procedures which are employed by a commander in planning, directing, coordinating, and controlling forces and operations."

To bring it down to a human level, a commander's life and that of his troops depends on swift, effective action based on a carefully managed and supervised pool of information regarding the position, location, movement, and status of all forces. Ultimately, decision making is the end result of the C^3I process. The acquisition, processing, and dissemination of data is the process by which decisions are made, and it is here that the computer steps in—to watch, to winnow, to warn.

The heart of the early electronic computer, a vacuum tube capable of completing 5,000 operations per second, has been replaced by solid-state circuits that today can handle 100 million operations per second. Experiments in advanced technology are

Artist's rendering of the sustained operation of AWACS, our Airborne Warning and Control System. (See AWACS surveillance plane in center of drawing.) A single AWACS is capable of tracking 400 aircraft simultaneously.

pioneering operations 50 times faster than those presently achievable. Experts are hard at work adapting commercial digital computer technology to military systems, a process largely involving "ruggedization"—the toughening of models to meet the exigencies of wartime environments. In the meantime, rugged or not, computers cope with the staggering amount of data available to the military decision maker and have an increasing say in who does what and when. (See page 137 for the role of artificial intelligence.)

WHAT C³I DOES

In essence C³I is a meshing of C² (command and control) *functions* with two sets of *systems* that allow the performance of C² tasks. C² functions include:

* Monitoring enemy troop strengths and resources.
* Monitoring U.S. and allied troop strengths and resources.
* Planning and replanning electronic warfare scenarios.
* Assessing warning signals and evaluating attack damage.
* Monitoring specific conflict situations.
* Choosing from among operational options and facilitating their execution.
* Assessing and controlling remaining military capabilities.
* Reconstituting and redirecting forces.
* Negotiating with the enemy and terminating conflict.

Operators at four computerized consoles aboard an AWACS surveillance aircraft.

The *systems* that make C^2 work are:

* Communications systems (the C that boosts C^2 to C^3) that insure *connectivity* with forces and data sources.
* Information-gathering and -processing systems (the I in C^3I), including computers, sensors, warning networks, and surveillance/reconnaissance devices.

It is noteworthy that, although it is only several years old, the concept of C^3I has already spawned an opposite war measure: C^3CM or counter C^3. *Counter C^3 is broadly defined as any action—such as the withholding of information through security, deception, jamming or destruction—that degrades enemy force control and delays its decision-making process.*

Dr. Gerald P. Dinneen, former Assistant Secretary of Defense for C^3I, says that to be successful, a C^3I system must develop in an evolutionary manner, since the various electronic systems that go into the total system are terribly expensive. It must also have the flexibility to provide our National Command Authorities and military commanders with a wide range of options. (Since requirements change and evolve over time, the use to which various sensors, data handling systems, and telecommunications are put will often be different from that for which they were originally designed.) Finally, a successful C^3I system must offer interoperability. The shifting strategic balance forces us to rely on the support of other countries, so that C^3I systems have to allow maximum interaction with our allies as well as among our own services.

Inside the North American Air Defense Command (NORAD) underground complex beneath Cheyenne Mountain, Colorado. In 1979 NORAD computers fouled up, putting SAC bombers and B-52s on enemy alert to counteract a false report that enemy missiles were attacking us.

The overall goal is a common, shared perception of what is happening at all possible levels.

Capabilities should be as simple and mobile as possible to allow for reconfiguration before, during or after attack. As terrestrial information systems become increasingly vulnerable, due in part to the Soviets' expanding arsenal of effective atomic warheads, the case grows for increased reliance on space-based C^3I systems. But such systems will have to be carefully "hardened" against growing Soviet antisatellite capabilities.

THE ROLE OF ARTIFICIAL INTELLIGENCE IN C^3I

In my view, we stand on a major threshold, one that offers great opportunity but is not without risk. We are ready to begin the rapid yet orderly transition of AI technology to military applications.

Rear Admiral Albert J. Baciocco, Jr.
Chief of Naval Research, 1980

Artificial Intelligence, or AI, is a branch of computer science that attempts to increase the reasoning and perceptual abilities of machine systems. AI is becoming increasingly linked to C^3I in two vital areas: information processing and decision making.

AI and Information Processing

AI technicians are working to find means for the efficient integration, storage, retrieval, and presentation of the literally billions of bits of raw data that feed into Command, Control, and Communication systems. Ultimately, according to Rear Admiral Baciocco, AI will free military personnel from typewriter ter-

Cutaway drawing of the underground Combat Operations Center of the North American Air Defense Command. Computers in such installations are now seen as "expert advisers," capable of assuming semi- or fully automated decision-making roles.

minals and allow them to interact with "intelligent machines" by:

* Generating messages in a simplified form of standard English.
* Speaking to the machine.
* Printing or writing requests on electronic tablets.
* Using touch screen displays.
* Drawing pictures and diagrams that are electronically disseminated.

AI and Decision Making

Computers are now seen as potential "expert advisers," capable of assuming semi- or fully automated decision-making roles. As part of this general thrust AI specialists are developing methodologies that will permit mechanical systems to "learn" from experience, adjusting their behavior to anomalies, inconsistencies, and change. "Control and negotiation concepts" and "situations assessment" are two other areas of AI research that are highly applicable to C^3I.

Baciocco has pointed out that the complexity of our own and enemy military networks and the speed with which tactical situations can change in an electronic warfare scenario mandate a need for the development of AI methodologies that will permit "dynamic task assignment." What Baciocco means is that *machines will decide who does what job and in what order.*

Another area in which automation will play an increasingly large role is *crisis alert.* Crisis warning concepts presently under study by AI researchers include:

* Supersmart systems with the initiative and "brains" to formulate and sound alert conditions on their own.
* Systems sensitive to the possibility of overalerting, thus capable of forestalling "cry wolf" situations.
* Systems that pick up and sort out inconsistencies or contradictions in alerts generated by different commands.
* Machines that can go beyond sounding a warning and can actually initiate a chain of "responsible" actions consistent with the nature of the crisis.

These last, it should be noted, will involve no interaction between the commander and the computer.

The machines will give the go-ahead for mission execution without human involvement.

EARLY WARNING AND C^3I

The nation's early warning network—a vital aspect of C^3I—is intended to insure that information regarding a potential ene-

The highly computerized missile warning center at NORAD. When computers present too much data, the decision-making process becomes obscured, creating what's known as information overkill.

my threat reaches the National Command Authorities in time to permit an effective decision based on a choice of options. The chief keystones in our early warning system include:

Apprehension of Enemy ICBMs (Intercontinental Ballistic Missiles)

* Three Ballistics Early Warning Systems radar stations extend their overlapping detection arcs 3,000 miles northward from sites in Greenland, Alaska, and Great Britain to search out missiles launched north from the Eurasian landmass.
* A Perimeter Acquisitions Radar Characterization System at Grand Forks, North Dakota, can track intercontinental missiles coming over the polar region from Siberia. It can also detect missiles fired from submarines situated in the Arctic and Hudson Bay regions.

Computer banks in the decision-making core of the underground NORAD complex.

In Andover, Maine, a permanent earth station for the Telstar satellite.

* Three space-based satellites, each equipped with 2,000 infrared sensors, "hover" above the Atlantic, Pacific, and Indian oceans. The Indian Ocean facility is equipped to sight rocket plumes of Soviet ICBMs within 90 seconds of blast-off.

By the late 1980s these satellites may be replaced by the High Altitude Large Optics (HALO) Program currently being developed by DARPA. HALO will do its detecting via sensors containing many computer chips, each of which will be capable of handling hundreds of thousands of "bits" or pieces of information. HALO's proposed 100-feet-wide optical panels, equipped with many of the tiny computer-

Pave Paws, a huge new satellite receiving station, was built by the Department of Defense.

ized sensors, should provide comprehensive and highly sensitive coverage.

Apprehension of Enemy SLBMs (Submarine Launched Ballistic Missiles)

* An FPS-85 phased array radar (see chapter 11 for description) at Elgin Air Force Base, Florida, protects against strikes from the Caribbean.
* Atlantic- and Pacific-based early warning satellites patrol these oceans for subs.
* Two powerful new phased array radar systems, called Pave Paws, situated at Beal Air Force Base in California and Otis Air Force Base in Massachusetts, are slated to scan the Atlantic and Pacific to a distance of 3,000 miles each.

Anti-bomber Systems

* Distant Early Warning (DEW line) radars situated in the Arctic wastes are being replaced by automated stations.
* Forty-five long-range radar stations jointly operated by the Federal Aviation Administration and the Air Force are being upgraded and automated.
* DOD is experimenting with Over-the-Horizon Backscatter radars (OTH-B), which are scheduled to become operational in the early part of this decade. OTH-B radars transmit signals that extend beyond line-of-sight from the ground, by bouncing signals to the ionosphere—about 25 vertical miles up. If the signals detect an airborne target, return waves reflect back to a receiver near the transmitter.
* A fleet of Airborne Warning and Control Systems—electonic surveillance planes referred to as AWACS—will beam word of potential attack. These modified Boeing 707s, which costs $131 million apiece, are equipped with IBM computers capable of tracking 400 aircraft simultaneously. The Boeings' Westinghouse pulse radar systems have a range of 350 miles at high altitude, and 250 miles at low altitude. Nine on-board consoles display the position of all targets, whether high- or low-flying.

* The Teal Ruby satellite, named for the mosaically patterned sensors it will carry, is expected by DARPA to be in orbit by 1983.

Comprised of a quarter of a million detectors, Teal Ruby will pick up the heat from jet engines.

WILL C^3I EVER REALLY WORK?

The C^3 business is not a simple one. It is an ill-defined and unbounded monstrosity. It is a necessary evil. It is an enigma with more people planning, programming, preaching, and engineering it than we have understanding it or effectively using it. We have more people buying and selling C^3 than actually producing it.

Vice Admiral Gravely
Former director of the Defense
Communications Agency, 1980

In a provocative article in *Signal,* the journal of the Armed Forces Communications and Electronics Association, in the spring of 1981, Roger Beaumont, a professor of history at Texas A&M, warned against the potential dangers of relying too heavily on C^3I as our savior in warfare. Among the problems he cited were the following:

There is no frame of reference that is shared by all the different individuals and agencies who participate in C^3I. Its members have generally worked with specific systems only, and have undergone specialized disciplinary training. Even the language of C^3I tends to be overly individual. Abounding in jargon and acronyms, it lacks a lexicon, or glossary. The resulting disparity of perspectives has led one observer to compare C^3I to the state of medicine in the Middle Ages.

Increasingly, the sense of human presence is missing. Tone of voice, timing, and changes in physical demeanor have long been significant aspects of command in battle—of running a war. In a defense state—or a war—that is managed by C^3I, interaction between staff and commander, especially in terms of personal and voice contact, is drastically curtailed. No more will a general spend up to eight hours a day in his jeep, traveling among his troops to personally bolster morale and instill confidence.

Though lack of human presence can be a disadvantage, it doesn't mean that machines don't have their place in modern warfare, says Beaumont. Command information is a fluctuating mass of uncertainties at best, and computers have a clear advantage over humans in their ability to organize uncertainty into a range of options at a far higher speed than can commanders and their staffs. But, he cautions:

"Reliance on mechanical aids undercuts the image of the commander as a person of superior intuition, judgment and skill."

Information overkill can result when computers present a mass of obscuring data. Obviously there is no set definition of essential command data and format. Although increased data flow in a time of uncertainty might *seem* to provide reassurance, in fact . . .

The usefulness of information may be inversely proportional to its volume.

Beaumont observes that the factual overload generated by the search for certainty may actually result in greater confusion.

Systems Dependency—or overreliance on systems—is an un-

healthy situation. Beaumont points out that the number of Americans who have been trained before the era of electronics, and thus have nonelectronic solutions to fall back on when these systems fail, is gradually dying out. Systems malfunction frequently. The United States must find ways of protecting our command system against the devastating effects of components going on the blink. This can and does happen, Beaumont points out, with unnerving frequency. (A 1978 report by the U.S. Comptroller General revealed that the supposedly refurbished C^3 system in the NORAD [North American Air Defense Command] Cheyenne Mountain complex could be out of the missile warning business for up to 15 critical minutes before backup equipment took over. A flagrant example of systems foul-up occurred in 1979 when NORAD computers in Colorado Springs mistakenly put SAC bombers and B-52s on enemy alert to counteract a false report that hundreds of enemy missiles were speeding toward us.)

Thus, Beaumont concludes, we must remain independent enough from machines to permit us to deal with penetration of the command network by the enemy, failure to report or misreporting of enemy tactics and arms, health and morale fluctuations among decision makers, acts of fanaticism and attempts at political gain. To counteract systems dependency, Beaumont recommends preserving ways to carry out coherent command-response operations with minimal or no use of electromagnetic wave radiation (EMR). Beaumont cites a number of alternatives that when listed together convey a strikingly old-fashioned sense of defense strategy, viz. signal flags, searchlight-Morse signaling, couriers, aircraft-mounted loudspeakers and sirens, rocket and artillery message relays.

Offering a final cautionary note, Beaumont points out that however elegant the superstructures of C^3 may be, the frequency with which complex communications networks are beset by both human and mechanical error even in peacetime raises a dim spectre for what may well happen in war.

The greatest risk of all would be to let our increasing dependence on C^3I's formidable electronic apparatus blind governments to the need for flexible policy-making and thoughtful negotiation in achieving our national goals.

RADAR REVIEW

Radar, an acronym for *radio detection and ranging,* is a system for obtaining, processing, and displaying positional information (on airplanes, meteorites, weapons), using electromagnetic waves. It was developed simultaneously by several countries during the 1930s. By the onset of World War II a continual 24-hour radar watch had been effected over Britain's main sea and air approaches.

Radars transmit electromagnetic energy in a particular direction. A target is detected by the reflection of that energy back to a receiver, which generally is located with the transmitter and shares a common antenna.

Radars are effective in all weather, night and day. Most are *pulsed,* meaning they transmit energy in the form of a pulse. This can happen at a rate of a few hundred to many thousands of pulses per second. Generally a new pulse is not transmitted until all returned energy from the most distant target has been received from the previous pulse.

In modern radar technology the computer calls the shots. Radar has come a long way since the days when tracks were hand-plotted. Today, thanks to computers, huge quantities of search and track data (see page 148) can be analyzed, correlated,

When launched, this aluminum contoured balloon will plot upper-level winds by radar track. The balloon is constructed in such a way as to give the best radar return.

and processed at astonishing speeds and converted to forms that can be readily understood. (The outputs of very modern radars are so complex that it is difficult for operators to understand them when they are displayed in raw video from. Therefore, computers are used to reduce data and produce pictures that show target positions and alphanumerics providing track number, height, and identity.)

In order for the computer to do its job, radar outputs must be put into numerical form. This conversion takes place in a device that converts the pulses into digits. As the digitalized information from air defense radars begin to stream into the computer, it tackles its first task—target declaration—converting information received into *plot data* (range, bearing, and elevation). A continuous stream of plot data is sent on to another part of the computer, which associates plots taken on successive scans and generates tracks. Once a track is established, the computer gathers information relative to that track and makes the correlated information available for display in the computer terminal.

The computer's next task may be to carry out a threat evaluation of each target tracked, based on an algorithm with which it has been programmed. Depending on target threat and position, it may next assign the target to a particular weapon system. The entire process—from target declaration to weapon assignment—can be carried out in milliseconds within the computer. Only when the fire order is about to be given may the human operator be advised, thus giving him the opportunity to intervene. Otherwise it's the computer that orders the appropriate fighter or missile into action.

In one system, the Phalanx, there is no human intervention at all. A close range sea-based weapons system, Phalanx consists of a radar, a computer, and a Gatling gun. Should a high-speed missile be detected speeding toward the ship, the computer automatically calls for fire.

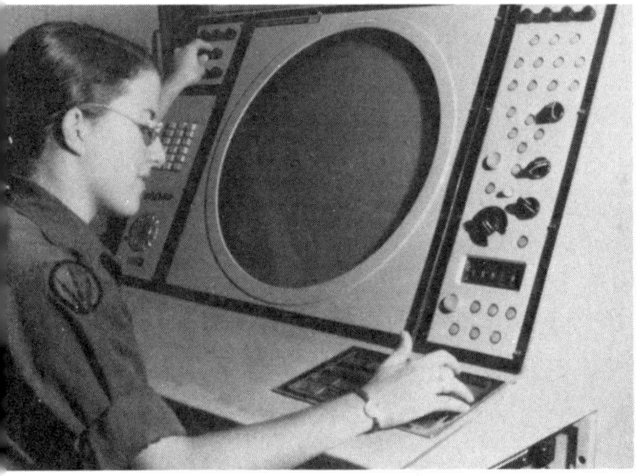

An instructor at the Army Air Defense Center in Fort Bliss, Texas, works with a radar screen.

There is no time to consult with a human. The computer, quite literally, calls the shots.

There are two basic types of radar: *search* and *track*.

Search radars are used to search out and detect the presence of targets in a large volume of space, such as over 360 degrees, out to 200 nautical miles, and up to 100 kilofeet. Typically, search radars can scan this volume in about ten seconds. Radars can be either 2D (giving position only), or 3D (giving position plus elevation). Search radars are used primarily in air traffic control systems and air defense systems.

Tracking radars track a particular target and obtain highly accurate and up-to-the-minute positional information. They remain locked on target by receiving on four beams. Two of these look left and right, two look up and down. By comparing the energy in the two beams of each beam pair, the radar can determine whether or not the target is moving away from the direction in which the radar is pointing, and the axis of the radar can be corrected accordingly. Tracking radars are used in weapons' fire control systems or for tracking satellites.

Threats to Radar

"Angels" Returns from unwanted targets such as ground, weather, and birds can create serious problems in radar communications. Some returns are difficult to identify and are called "angels." These false targets can be eliminated or greatly reduced by moving target indicators (MTI), circuits that measure the velocity along the radius of targets by Doppler effect and filter out those targets that are stationary or are proceeding very slowly. (The Doppler effect is defined as the change in the observed frequency of sound, light, or other waves incurred when either the observer or the source of the waves shifts position. A classic example is the change in the pitch of a train whistle as the engine passes by.)

Jamming Military radars are vulnerable to jamming. Jamming can be active or passive. An active jammer generates a signal at the radar frequency, generally in the form of noise but sometimes simulating other targets. Passive jamming occurs when metalized strips, or chaff, are dropped from aircraft to confuse radars. Experts are hard at work developing electronic counter-countermeasures to maximize the effectiveness of C^3I networks in the face of such enemy interference.

SONAR AT A GLANCE

Sonar, an acronym for *sound navigation ranging,* uses underwater sound to detect the location of objects in the sea.

In the early 1900s French scientist Paul Langevin pioneered an ultrasonic water device that sent bursts of high-frequency sound waves through water. His work forms the basis for modern sonar technology, which uses echoes from ultrasonic waves to detect targets.

Sonar works much like radar except that it moves more slowly through water by several orders of magnitude than radar travels through space. Since electromagnetic radiation such as radar and visible light have a limited ability to penetrate water, sonar is widely used in naval warfare to track submarines. Sonar is also extensively employed by submarines, usually in the "passive" form (see page 150), since the submarine's chief aim is to remain undetected by running quiet. (Linkage between submarines and communication networks is currently being improved through research in extremely low frequency [ELF] radiowaves, which have good penetrating powers. In spite of protests from environmentalists and others fearing low radiation effects, the Reagan administration recently launched an ELF program which involves sinking 84 miles of antenna wire in Michigan and Wisconsin for communication with oceangoing subs. The U.S. Navy allots $10 million a year to research the nature of sound and how it moves through the sea.)

There are two types of sonar, *active* and *passive.*

> *Active,* or echo-ranging sonar, involves a pulse of ultrasonic waves generated by a transmitter being projected into the water. When it contacts the target, the pulse is reflected back and picked up as an echo by a receiver. After calculating the speed of sound in seawater (a function of water temperature), the range of the target can be determined by measuring the length of time it takes a sound wave to be transmitted to the target and back. (At 25 degrees Celsius, for example, sound travels at the rate of 1,531 miles per second.) It is also possible, either by analysis on a cathode-ray tube or simply by careful listening to roughly classify targets (submarine contact, school of fish, wreck, etc.) and determine whether the target is advancing or receding.
>
> *Passive* sonar systems permit transducers to pick up noises emitted by targets or by their machinery as they move through the sea. Since every individual ship has its own characteristic noises, passive sonar helps classify targets. It is used mainly by submarines to detect surface vessels.

Militarily, sonars are used with minesweepers (to locate mines); by helicopters (which detect submarines by lowering "dunking" sonars on cables); and by acoustic torpedoes (for homing in on targets).

Nonmilitary uses of sonar include locating schools of fish, locating wrecks, monitoring fetal development and other medical applications, and finding the depth of water under a keel.

Seaborne sonar devices can be towed through the water by complete electronic packages called "fish" or dropped into the sea from aircraft as sonobuoys. Sonobuoys are tiny expendable broadcasting stations that receive water sounds and transmit them to aircraft. Upon hitting the sea their antennae are erected and a microphone is lowered to suitable depth. Power for their electronic circuitry is provided by seawater streaming into a water-activated battery.

The problem with sonar is that propagation of the ultrasonic waves is greatly affected by the characteristics of seawater itself. Ultrasonic waves will bounce off boundaries between different temperature layers in the water and be lost. Since no amount of power in the sonar beam can penetrate these temperature layers, submarines will hide below them. One solution developed so far is the "variable depth towed array," a sonar system that is lowered over the stern of the ship until it is below the obstructing layer.

Sonars and Computers

Computers perform a variety of functions in the area of underwater detection. According to the Stockholm International Peace Research Institute (SIPRI) yearbook, the U.S. Navy is isolating and taping individual submarine noises using its Illiac 4 computer at the California-based Ames Research Center. The tapes are sent on to the U.S. antisubmarine fleet to be played and compared with submarine noises picked up by the sub's own sen-

Artist's rendering of new Defense Department radar receiving stations.

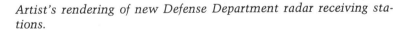

sors, better to pinpoint the identity of potential underwater targets.

THE NEW CRYPTOGRAPHY: THE CODE ROOM GOES PUBLIC

A safe system of code making and breaking is essential to effective C^3I. Until recently U.S. security experts could breathe easy, certain that the latest advances in cryptography were carefully guarded behind the high fence surrounding the top secret National Security Agency (NSA) at Fort Meade, Maryland.

Today the situation is changing as both academe and industry begin to focus on independent cryptological research, challenging NSA's monopoly and creating what some see as a serious conflict between national security and scientific freedom. Indeed, NSA's fears were fanned five years ago when a new type of cryptography was devised by three Stanford researchers. Called "public key cryptography," the system is intended primarily for business applications. It is called "public" because the key to the code can be transmitted openly.

What is making cryptology, formerly a discipline reserved for the cloak-and-dagger set, go increasingly public?

The new interest evidenced by academic scientists in cryptological research has a theoretical basis that is linked to fundamental mathematics. In the case of industry the motivation is strictly practical: safe codes are needed to protect computer-stored commercial information against theft and to guarantee the authenticity and confidentiality of transactions executed by electronic means. (The threat of computer theft has become so great that Lloyd's of London calls it "potentially one of the fastest growing fields of international crime" and in October 1981 issued its first policy to insure a company against electronic and computer crimes.)

NSA is alarmed by the possibility that cryptography papers appearing in scholarly journals might leak new coding tricks into hostile hands, enabling unfriendly governments to decipher the NSA's codes and permitting them to harden their own codes against cracking by U.S. experts. Moreover, the NSA would undoubtedly want first look at any code devised by an American that proved more ingenious than those in its own repertoire.

In an attempt to maintain control over cryptography research, NSA has taken a number of controversial steps:

* On August 10, 1980, it prodded the National Science Foundation into withholding funds for portions of a cryptography research grant awarded to MIT's Leonard Adleman, which the agency felt might endanger national security. This is believed to be the first time the NSF refused a researcher funds for reasons unrelated to the merit of a proposal.
* In October 1980 it persuaded the Public Cryptography Study Group, a nine-member panel, most of whom represent professional societies in computer science and mathematics to approve a two-year voluntary system of prior restraints on research publication in cryptography. If at the end of that time the voluntary system had not proved "useful and efficient," NSA was to be able to get legislative authority for mandatory restraints. To date 30 papers from MIT have been submitted to the agency for review.
* In August 1981 it formed a group designed to coax companies into sharing their work on computer security with NSA personnel.

Such NSA actions have raised backs in academic and industrial circles. Although none of the MIT papers vetted by NSA to date have met with objection, George Davida, a computer scientist at the University of Wisconsin in Madison, maintains: **"The agency's attempt to control academic research is harmful to the health of cryptology research and to the health of our civil liberties."**

One shot of an air-to-air missile can cost more than most men will earn in a lifetime.

chapter 11

The Spiraling Cost of Megaweapon Defense

The administration has not yet told us how it plans to spend the record $160 billion it allotted in 1980 for defense—25 percent of the entire federal budget. The public only knows it plans to spend a great deal, increasing allotments yearly until the defense budget reaches $367.5 billion—or $1.3 trillion—in 1986. This constitutes the largest peacetime military buildup in American history. Over the next five years military expenditures will be nearly three times the cost of our most recent war. In 1980 dollars Vietnam cost a modest $59 billion.

One reason so much money is going to the Pentagon is that the new weaponry the Department of Defense wants is spiraling in cost. The Pentagon has conceded that the price tag for 47 major weapons-programs soared by $47 billion during the final three months of 1980 alone. Among the new MEGAWEAPONS whose actual costs rose drastically from what was originally budgeted are:

* The Navy's F-18 fighter plane, which rose by $8.1 billion.
* The Army's XM-1 tank, which rose by $5.9 billion.
* The Army's new infantry carrier, which rose by $5.3 billion.
* The Navy's nuclear-powered Trident submarine, which rose by $2.9 billion.

THE RELATIONSHIP BETWEEN WEAPONS COMPLEXITY AND WEAPONS COST

One of the most striking examples of how the cost of gargantuan weapons-design projects can get out of hand is the Trident submarine, which is longer than the Washington Monument is tall. Carrying 24 missiles, it was meant eventually to replace the 41 Polaris and Poseidon submarines, each of which carries only 16. Nine Tridents have been authorized, budgeted at a cost of $900 million each.

The first Trident submarine, which was to have been put to sea no later than April 1979, finally went into service in June 1981, more than two years behind schedule. Its price tag had hit $1.2 billion. The cost overrun and delay are generally chalked up to extensive repairs and reworking necessary to make the gargantuan sub seaworthy, but everyone involved passes the buck. The Navy blames the sub's builder, General Dynamics Corporation. Company executives insist that the Navy furnished defective equipment and issued numerous design changes—more than 2,900 in one three-month period alone. Now the Navy wonders whether, at $1.2 billion a shot, it can afford the planned Trident fleet of 14 or more. Nuclear-armed submarines are supposed to be the least vulnerable of our nuclear system because theoretically

Close-up of a Sidewinder missile installed under the wing of a Tomcat fighter aircraft. One of these missiles costs over $10,000.

they could survive a first strike and retaliate. But because of the delays building the Trident, **the Navy has fewer nuclear armed submarines at sea than at any time since the 1960s.**

This kind of fiasco is characteristic in all weapons work, regardless of which branch of the military is responsible for it. When the Army teamed up with the Chrysler Corporation to build the new XM-1 tank, they knew the Russians produced twice as many tanks as we do, so they tried to cancel that advantage by building a supertank. The resultant XM-1 is the fastest, most sophisticated tank ever built. But it is also the heaviest. It can't go nearly as far as our old tanks. And, in an oil-short era, it uses 1.9 gallons of gas per mile. The XM-1 tanks ended up costing $2.7 million apiece, even though the original version of the tank, designed in 1963, was supposed to cost $420,000.

It has taken 18 years and 3 separate designs to produce a tank that was considered urgently needed in 1963. In the end the entire project—including false starts—will cost more than $21 billion.

All along, the Army has been planning to build 7,000 of these tanks, but at what they've ended up costing, it isn't clear that we'll be able to afford the full battalion.

Every major American weapons system today is the product of the same inherently faulty thinking: design and build a few highly complex megaweapons instead of having to build many simpler ones. Each generation of planes, tanks, missiles, or ships, ends up costing between two and ten times more than the previous one. And the designs themselves often broach the ridiculous. In his book *The High Cost of Defense,* James Fallows satirizes the way the Pentagon fantasizes in coming up with its defense concepts.

Wouldn't it be great to have a tank that went from zero to sixty like a hot rod? What if we had a missile with a 99-per cent 'probability of kill,' or one that would give American forces a 955-to-one exchange rate against Soviet aircraft? What if we could build an electronic fence across Vietnam and Laos to keep all the Communists out? Wouldn't it be wonderful if, instead of leaving aerial combat to a group of pilots trying to figure out for themselves which enemy planes to destroy, the whole enterprise could be automatically controlled from the ground?

"The way it works is pure salesmanship on the lab's part. I call it educating the only customer in town. We come up with a new bomb. The lab's sales office, the Military Applications Program, writes up the specs and sends one of its salesmen to the Air Force, Navy or Army. The salesman says: 'General, wouldn't you like a new bomb for the B-1?' If he says yes, the specs then go from the Pentagon to the Joint Chiefs [of Staff], who ask the National Security Council, which makes a recommendation to the President, who issues a directive to the Department of Defense, which orders the bomb from the Department of Energy and the lab. Defense gets a new bomb and the lab has the prestige of getting its weapons in the stockpile."

A veteran Livermore scientist,
as reported in *The New York Times Magazine*,
January 18, 1981

It is easy, says Fallows, to see how the complicated MX missile system was dreamed up.

Wonder weapons seem the perfect solution—they proffer the seductive idea that warfare can be conducted easily and automatically, when in actual fact it is always confused and messy. We know that the Soviets build more weapons than we do, and that they go in for numbers rather than sophistication. Since So-

"Sure, it requires some sacrifices, but I think the nation can afford it."—Frank C. Carlucci, Deputy Secretary of Defense. (Statement made in a military budget meeting calling for congressional authorization of $258 billion, February 1982.)

viet weapons outnumber ours, we have turned to our superior high technology. "Given our disadvantages in numbers," former Secretary of Defense Harold Brown has said, "our technology is what will save us." But often the weapons are too delicate or impractical to be of use.

Every year hundreds of officers and other weapons specialists retire from the government and go to work for private defense contractors. By the same token hundreds of men from the weapons industry leave their corporate jobs and go to work in the Pentagon. "Here," a recent CBS report on defense noted, "the revolving door is not considered a conflict of interest. It is simply a fact of life. And it's also a fact of life that as buyer and seller merge, perspectives narrow, alternatives disappear, and debate stops." It's assumed that a Pentagon official who wants a second career in the defense industry will not go overboard in criticizing a contractor's cost overruns, nor will an industry executive eyeing a Pentagon position be quick to point out that a new weapons design is preposterous. In Fallows's view, this revolving-door situation between industry and the government creates a "culture of procurement" in which real defense needs get lost in the game of creating jobs and keeping the economy going. The military, he says, ends up "getting drawn toward new weapons because of their cost, not in spite of it."

ECONOMIC RESULTS OF DESIGN COMPLEXITY

The Army's latest tank, the XM-1, costs at least seven times as much as the Sherman tank of World War II. The Main Battle Tank—a proposal from the early 1970s that was junked because of technical problems and resistance from Congress—would have cost ten times more than the Sherman. Modern aircraft carriers, with their nuclear propulsion, cost four to five times as much as those built at the end of World War II. The first guided missile (used for aerial dogfights and still the most reliable) is the heat-seeking Sidewinder, which originally cost about $3,000 apiece—about $10,000 in today's dollars. Its newer and less reliable radar-guided counterparts, the Sparrow and the Phoenix,

> Aside from the 2.1 million uniformed personnel and 850,000 reservists who currently make up our armed forces, and the additional one million civilians who work for the Department of Defense, 35,000 of them in the Pentagon itself, there are today more than 1,100 major American corporations employing a total of more than 700,000 workers, and engaged almost exclusively in the research, design, development, testing, evaluation and production of weapons. This complex web of government agencies, private organizations, think tanks, business firms, bureaucracies and lobby groups has spun about itself so many threads of mutual affirmation, shared assumptions about the national interest, common policies on parity, deterrence and readiness, and underlying philosophies of human behavior as to weave an almost impenetrable fabric of doctrine to validate its vested interests.
>
> Tom Gervasi
> *Arsenal of Democracy*

cost ten and one hundred times as much respectively. The latest model of the Phoenix costs more than $1 million.

Higher and higher costs actually degrade military readiness. When so much money goes to fancy computer circuitry and complex engine designs, there is less available for the meat-and-potatoes of defense readiness: training, extra rounds of ammunition, realistic preparation for combat. What happens is that we end up with fewer weapons that work less well. A civilian employee of the Pentagon's Program Analysis and Evaluation Office, Franklin C. Spinney, has spent the past few years studying the implications of the trend toward complex weapons. The Spinney Report, officially entitled "Defense Facts of Life," was made public early in 1981 through the efforts of Senator Sam Nunn of Georgia, one of the fiercest hawks on the Senate Armed Services Committee. Spinney concluded that rising costs and complexity have had an immediate effect on the number of weapons we have. Each weapon can be used less often than its simpler, cheaper alternative. **As an airplane or missile becomes more complicated, the probability that all its parts will be working correctly at the same time decreases.** "Our strategy of pursuing ever-increasing technical complexity and sophistication has made high-technology solutions and combat readiness mutually exclusive," Spinney claims.

One example can be found in expensive and unreliable aircraft, which dramatically reduce what fighter pilots call "presence in the sky." As Everest Riccioni, a retired Air Force colonel now working at Northrup, points out, there's a difference between the "real fleet" and the "phantom fleet." The real fleet is the one that can be put in the air at any given moment; the phantom fleet is composed of all the planes that are confined to the runway for repairs or lack of crews. Says Riccioni . . .

"The enemy doesn't give a damn about the phantom fleet. The only force that matters to him is the one in the air."

Six years ago the Navy and the Air Force organized a simulated combat test to try out F-14s and F-15s: big, expensive planes with long-range missiles. One side flew F-14s and F-15s, the other side flew Northrop F-5s. The U.S. military considers the F-5 too small and crude for anything other than sale to allies, but it was used in the tests because of its similarity to the Russian MiG-21. To the shock of the Navy and Air Force the big, missile-carrying planes lost out consistently to the smaller planes because once airborne, they made far more visible targets.

If there could be a demonstration of the principle that more spending leads to more and better defense, one would expect to find it in the story of Tac Air (Tactical Air Command). But the dismal truth is that ever-increasing monies spent seem to have accomplished just the opposite. F-14s and F-15s form the Air Force's Tactical Air squadrons, consisting of all those planes that are not designed for strategic nuclear asssault on the Soviet Union. They include fighter planes, attack or bombing planes, radar planes, reconnaissance planes, radar early warning planes, electronic countermeasure planes. Since the early 1960s the Tac Air share of the Air Force's budget for new equipment has jumped from 21 to 62 percent. The aircraft engines need many more exotic materials to withstand high internal pressure and temperature.

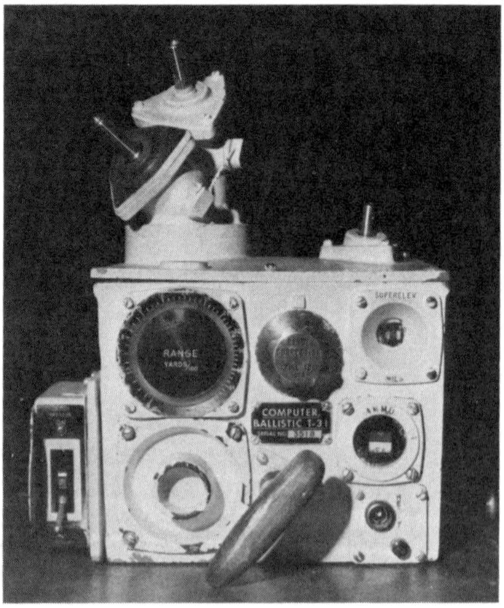

Ballistic computer for the M-60 tank. Computer repair causes a lot of "down time" in the modern military.

The planes' frames need expensive materials like Titanium, stainless steel, beryllium. The fighter planes have large radars to detect enemy planes and computerized systems to guide missiles to their targets. (F-14s and F-15s are theoretically able to locate and destroy targets at night, and in rain and fog.) But each of these new developments has pushed up the planes' costs inexorably. The F-15 costs between $26 million and $35 million each, the F-14 between $15 million and $25 million. According to Norman Augustine, a Martin Marietta Aerospace vice-president who used to work at the Pentagon, "From the days of the Wright brothers through the F-18 [a current Navy fighter], aircraft costs have been increasing by a factor of four every ten years.

"If the trend continues, in the year 2054, the entire defense budget will be able to purchase just one tactical aircraft. This aircraft will have to be shared between the Air Force and the Navy, three and a half days each per week."

THE PROBLEM OF REPAIR

Our vaunted F-15, Spinney reports, is "non-mission capable" 44 percent of the time. Partly this has to do with bunked up computers and the problems involved in repairing them. Each F-15 contains a total of 45 "black boxes"—computer systems that run the avionics and other functions of the plane. When one of the boxes malfunctions, the whole thing is removed and a new one snapped in. The defective box is sent back to a depot for servicing. To identify the defective part, the maintenance crew requires an extremely expensive and very complicated computer called an Avionics Intermediate Shop. The Shop supposedly simplifies and speeds up flight line maintenance. But in 1979, says Spinney, the system only worked 50 percent of the time; the following year it improved to 80 percent. Even when the Shop is working, it may not find anything wrong with a box pulled from a plane. Then the box is put back on the shelf, to be tried later in a different plane.

If the Shop does track down an error—a maddeningly lengthy process—it may need an average of three hours, and sometimes up to eight, to check one box. And the Shop can only check one box at a time. Spinney rounds out the bleak picture by explaining that if and when the Shop itself goes out of commission, an entire F-15 squadron is left without maintenance.

Another area of defense in which computer repair becomes a crucial issue is the military's massive radio and computer hookup designed to control military maneuvers from one central point. This Worldwide Military Command and Control System, known as WWMCCS, or Wimex, has had over $10 billion invested in its design and upkeep. When the system was thoroughly tested in 1977, attempts to send messages ended in breakdown 62 percent of the time. In one 18-month period there were 147 false

An Army gunner shows how to fire a TOW missile system from the ground on a lightweight launcher mounted on a tripod. Presumably this missile, which costs about $10,000, can destroy a Soviet tank worth a million and a half. But the infantryman, exposed on a field, must take unusual risk with his life. Some defense experts believe soldiers would abandon TOWs in time of battle.

alarms triggered by such things as a Soviet training launch, a failed computer chip worth 64 cents, and a gas fire on the Siberian pipeline. Centralized command networks are useless when they don't work, and may be actively harmful—bringing us a hair's breadth from accidental nuclear war—when they do. (See chapters 10 and 17.)

Also functionally unreliable are missiles designed to "kill" tanks on the battlefield. The Maverick missile is supposed to be able to "see" different-shaped targets and select the right one to hit. If it "sees" a tank, a bridge, and an armored personnel carrier and is programmed to hit tanks, it will "choose" a tank as its target. But sometimes the Maverick is confused. In tests, it has attacked telephone poles, fenceposts, rocks. The Navy is working on a new improved Maverick, which will probably cost twice as much as the present price of $70,000 each.

Another antitank missile, the TOW (tube-launched, optically tracked, wire-guided), has serious battlefield problems owing to the complexity of its design. A soldier has to stand up on the open field and fire the missile at a far-off tank. By keeping the sight on the launcher aligned on the tank during the ten seconds or so of the missile's flight, he can adjust its course as the tank maneuvers, so that a missile that costs $7,000 can presumably destroy a Soviet tank worth $1.5 million. But in the meantime he has taken an excessive chance with his own life. As one planner says, "The people who think these things up don't realize there's all the difference in the world between a two-second exposure on the battlefield and ten seconds." In designing TOW its planners seemed to have concentrated on TOW's "probability of kill" to the exclusion of such practical factors as the gunner's chance of survival and the rate at which he is able to fire. Soldiers will often discard weapons they don't have faith in, and some defense experts have suggested that TOWs will litter the first battlefield where they're actually used.

FOUR MAJOR SPENDING FALLACIES

Unrealistic military planning, which has ignored the economic and military effects of complexity, constitutes "a form of organizational cancer," according to Spinney. The disease progresses in four main steps. First, the planners assume that Congress will give them ever greater sums of money to spend on new equipment. Second, they forget or fool themselves about the amount of money they need for maintaining their complex new equipment. So—the third step—the military has to make unexpected cuts in its investment budget (the money set aside for new equipment) in order to make up the cost of maintenance and overruns on previous weapons. Thus, we end up buying a smaller number of new weapons because we don't have the money to buy more. Fourth, when the military's economy is in good shape, it doesn't use its money for bailing out projects it has already launched, but to buy yet another complex weapon, with yet another inadequate maintenance system. "In a general sense," Spinney says . . .

"This pattern reflects a tendency to reduce our current readiness to fight in order to modernize for the future."

Future plans for new weapons seem to repeat the same kinds of errors in thinking Spinney describes. Increasingly complex systems are in the ascendancy. DOD has invested heavily in making "smart" (electronically guided) weapons even smarter, through miniature circuitry. All precision-guided missiles like

Norman Augustine, vice-president of defense contractor Martin Marietta, says, "Aircraft costs have been increasing by a factor of four every ten years."

the Maverick, for example, depend on sensors to home in on their targets, or to direct them to follow a desired path. The sensors depend on making mechanistic yes/no comparisons to distinguish the target from its surroundings. Infrared sensors choose between hot and cold; radar guidance measures the relative strength and frequencies of waves reflected from the target and the background. Some systems use television cameras to search for black/white contrasts in the target images.

If this equipment sounds as delicate and unreliable as the computer systems in the F-15, it is: when the "smart missiles" went to Vietnam, most of them were flops. One set of missiles whose vaunted "probability of kill" was 99.5 percent actually worked only 7 percent of the time. It cost $2 billion. Yet Norman Augustine of the weapons contractor Martin Marietta promises that by decade's end, the computational power of missile computers will actually be close to that of the human brain. "Our missiles will not be just smart, they'll be intelligent."

Yet, for all their "intelligence," our expensive new megaweapons seem never to remain immune to countermeasures. For example, the most highly touted prospect for military high technology is millimeter-wave radar guidance that supposedly will enable each missile to distinguish a tank from a car and an artillery piece from a rock. But corner reflectors—devices that send back disproportionately strong radar reflections—could easily fool the new missiles. One Pentagon official envisioned mounting corner reflectors on "a bunch of motorcycles going the other way. You'd see the missiles take off after them." As for plane radar, F-15 and F-14 pilots complain that they get three good solid flights or a total of about three and a half hours in the air before their radar breaks down completely.

Fallows's point about "the corruption of military purpose by procurement" is that the goal of the current system seems to be to spend money, not figure out whether a weapon works. A new weapons system is rarely canceled. The procurement system insures that a weapon will have a life of its own, regardless of how it performs out in the field. There's no profit-and-loss sheet in the Pentagon—and thus no competition and incentive to produce better products. The military machine, in substituting procurement for defense, has lost its sense of purpose.

U.S. Cobra helicopters being prepared for shipment to Europe.

chapter 12

Dealing: How the United States Profits from the Sale of Arms

When President Eisenhower left office in 1961, he sounded the first serious alarm about "the military-industrial complex," the Iron Triangle composed of big defense contractors (Boeing, Lockheed, McDonnell Douglas, etc.), procurement officers in the Pentagon who often retire from the military to take jobs with the same arms companies they used to buy from, and senators and congressmen who belong to the congressional committees that authorize big sums for military spending. Because of the Cold War, said Eisenhower, "We have been compelled to create a permanent armaments industry of vast proportions. This conjunction of an immense military establishment and a large arms industry is new in the American experience. The total influence—economic, political, even spiritual—is felt in every city, every state house, every office of the federal government."

Ike was worried about domestic politics, specifically, the growing numbers of former military officers who were working for arms companies, and the threat that "tremendous peacetime military spending" might unbalance the economy. But he might have been referring as well to the effect of unrestricted arms dealing on our foreign relations.

The international arms market is increasingly competitive, as ever more countries vie to sell arms abroad.

The United States has maintained a comfortable lead, selling more than half the $22 billion traded in weapons every year.

(Its closest competitor, the Soviet Union, accounts for 24 percent of sales, followed by France and Great Britain.) Since 1978 there has been a record number of new sales of almost every weapon produced in the United States. Most of the selling is done by the Pentagon, which in the past eight years has moved $79 billion worth of U.S. military equipment and services to foreign buyers. American military advisers in foreign capitals often find themselves acting as Pentagon sales reps.

The pressures of corporate economy have contributed greatly to dangerous arms trade. Like other manufacturers, companies that make arms for the Pentagon plan for bigger profits every year, regardless of whether there's a need for the product. In the early 1970s Lockheed, then the Pentagon's number-one supplier, faced bankruptcy and had to be bailed out by federal loans. The company began to look overseas for new markets and new profits. Arms companies' quest for profits is probably the chief reason for the burgeoning arms business, although the government itself has been eager to recycle petrodollars. Most American arms sales go to oil-producing countries, and to Israel.

These economic pressures have been helped along by political decisions, notably the Nixon doctrine. In an attempt to avoid future wars like the one in Vietnam, Nixon and Kissinger thought it was a good idea to send American arms rather than American personnel into unstable areas. The plan often backfired. In the belief that Iran had the power to stabilize the Middle East, Nixon assured the Shah in 1972 that he could buy anything he wanted from the American arsenal. Before his fall from power the Shah bought more than $20 billion worth of American arms. In 1980, when we attempted to rescue our hostages in Iran, we ran the risk of being brutalized by the very weapons we had sold the country. Subsequently, Saudi Arabia has replaced Iran as our biggest arms client: the hope now is that *it* will be a stabilizer, even though its hostility to Israel threatens to effect just the opposite result.

THE ARMS MERCHANTS

Weapons Sales to the Third World*

	1974	1975	1976	1977	1978	1979
Total	23,521	22,329	21,394	27,356	24,198	29,978
Noncommunist total	16,581	17,979	14,254	17,606	20,458	19,258
United States	11,921	11,614	10,669	9,976	11,268	10,388
France	2,030	2,300	1,025	2,800	2,500	4,000
Great Britain	760	1,400	630	1,550	1,800	2,420
West Germany	725	790	360	1,170	2,220	400
Italy	425	990	220	960	1,360	360
Other	720	885	1,350	1,150	1,310	1,690
Communist total	6,940	4,350	7,140	9,750	3,740	10,720
Soviet Union	5,900	3,600	5,900	9,000	2,900	9,800
Other	1,040	750	1,240	750	840	920
Dollar Inflation Index (1974 = 100)	100	109	118	127	136	148

*Prices include sale of weapons, construction, military assistance, and spare parts. Third World category excludes Warsaw Pact, NATO countries, Europe, Japan, Australia, and New Zealand.

As recent events in the Middle East have proven, the Nixon doctrine isn't reliable. There's no way to guarantee that a country whose arms we supply won't turn against us. Nor can we be sure that a buyer will fight the wars we want against the adversaries we've chosen for them. When India and Pakistan went to war in 1965, and when Israel attacked Jordan in 1967, both sides used American weapons. We hope the weapons we sell will be deterrents, but sometimes they simply make it easier to go to war.

Arms are unique commodities. It's difficult to measure to what degree they actually dictate policy. Did Nixon come up

Types of Weapons Delivered (1973–79)

	United States	Soviet Union	Western Europe
Tanks and self-propelled guns	7,007	12,565	2,395
Artillery	4,341	5,675	975
Armored personnel carriers and armored cars	14,071	10,545	3,425
Major surface ships	89	7	24
Minor surface ships	162	135	264
Submarines	19	9	24
Guided missile boats	0	82	30
Supersonic combat aircraft	1,452	2,950	475
Subsonic combat aircraft	924	580	57
Helicopters	1,352	940	1,500
Other aircraft	973	385	945
Surface-to-air missiles (SAMs)	8,935	19,495	945

with his doctrine because the arms companies needed new markets, or did the new markets follow upon the policy? Unlike other salesmen in search of new markets, all an arms salesman has to do is to sell a new product to one country and then wait for its neighbors to become new buyers hoping to protect themselves. With new types of weapons continually rendering old ones obsolete, regional arms races are maintained by a more or less chronic state of panic, just as is the race between the superpowers. In June 1981 the United States announced that it would sell advanced F-16 fighter jets to Venezuela and Pakistan—virtually guaranteeing that their neighbors will also start buying from us. And both Taiwan and the People's Republic of China are now customers for U.S. arms.

Because arms companies depend heavily on huge, one-shot government contracts, the arms business has always been associated with a delicate potpourri of bribery, intelligence activity, and diplomacy. In Europe after World War I there was widespread public revulsion against the "merchants of death," who were held responsible for fanning conflicts in order to promote sales. The most famous arms dealer was Sir Basil Zaharoff, agent for the British firm of Vickers (now Vickers-Armstrong), who func-

Developing nations used to complain that if we refused to sell them arms we were keeping them defenseless, and dependent upon us for protection. Now they argue that by selling them arms we keep them equally dependent upon us because we keep them poor. They are right. It has been argued that they don't have to buy. Neither, it should be countered, do we have to sell.

But of course we believe that we must sell. We believe this because we believe arms are needed. We believe they are needed because we are making them.

Tom Gervasi
Arsenal of Democracy

tioned as a salesman and a spy. He was not above promoting wars or using bribery to achieve his ends. Before World War I his letters to the home office spoke of "doing the needful in Portugal," "greasing the wheels in Russia," "administering doses of Vickers to Spanish friends." In the 1930s he tried to "ginger up Chile." "I sold armaments to anyone who would buy them," Zaharoff reminisced to a London journalist in 1936.

"I was a Russian when in Russia, a Greek in Greece, a Frenchman in Paris. I made wars so that I could sell arms to both sides."

In 1975 a Securities and Exchange Commission (SEC) and Senate investigation of Lockheed revealed that little had changed in the arms business since the greedy machinations of Sir Basil. Lockheed, the investigators found, had paid out $200 million as commissions and agents' fees. At least $38 million had gone to foreign officials, including Prime Minister Kakuei Tanaka of Japan and Crown Prince Bernhard of the Netherlands. The repercussions overseas were immense: Tanaka fell from power and the Dutch government was badly shaken. Lockheed's president and chief executive resigned, while the company defended its payoffs as "consistent with practices engaged in by numerous other companies abroad." In 1979 the Justice Department dropped plans to prosecute Lockheed's former president, apparently because he planned to defend himself by maintaining that the federal government—notably the CIA—was aware of the payoffs.

Since some arms companies supply the CIA, the links between arms dealing and the intelligence community probably constitute a miniature military-industrial complex. The Wilson-Terpil case is a good example. Several years ago two ex-CIA operatives, Edwin P. Wilson and Frank E. Terpil, struck a lucrative deal with Muammar al-Qaddafi, Libya's ruler, who supports terrorist networks in the Middle East, Europe, and Africa. According to another former former CIA agent, who worked for Wilson and Terpil, the two used the cover of their import-export business and their CIA contacts and experience to help Libya set up a plant to manufacture assassination weapons, to help Qaddafi plan political assassinations, to recruit former Green Berets to teach Libyan soldiers and Arab terrorists how to handle explosives, and to smuggle arms and explosives to Libya. Wilson, who has made millions from his arms scheme, set up his own company in Tripoli to supply Qaddafi's army.

Wilson and Terpil's former employee, who at first had thought they were running a CIA front, blew the whistle in the fall of 1976 when he was told to procure a Redeye ground-to-air missile, a heat-seeking missile that can destroy aircraft in flight and that could not be legally exported from the United States to Libya. The agent was afraid that if Qaddafi wanted only one missile, it meant he was planning to use it in a terrorist attack, possibly for shooting down a 747.

"I felt that Frank and Ed were giving Qaddafi any goddamn thing he asked for."

No charges were brought against Wilson and Terpil until 1981, which gave them plenty of time to flee overseas. They also, it appears, sold arms and expertise to South Africa, Uganda, and DINA, the Chilean intelligence agency that assassinated former Chilean leader Orlando Letelier in 1976 on American soil. And some agents still with the CIA collaborated with Wilson and Terpil to make money illegally. The pair may have used agency connections to obtain Pentagon and National Security Agency defense purchase plans, which they then sold to a corporation. Wilson, who has many friends in high places in Washington, as well as a great deal of knowledge about how U.S. Intelligence works, may still be enjoying CIA protection. The agency has denied that it had any official connection with Wilson and Terpil, but they could have been free lancers whom the CIA was using to manipulate—and even stimulate—terrorism. Terpil told *60 Minutes* that the CIA knew all along what he and Wilson were up to in Libya.

The Terpil-Wilson case highlights the glaring deficiencies that exist in the laws and policies governing the sales of American arms and technology overseas. Incredibly, Terpil and Wilson could be convicted of smuggling, but not for planning assassination. There are no U.S. criminal laws that prohibit American citizens from training terrorists overseas, or from selling arms expertise to hostile countries. Qaddafi wasted no time benefiting from Terpil and Wilson's instruction. He has used their matériel and know-how to assist the Palestine Liberation Organization (PLO), the Red Army of Japan, the Irish Republican Army (IRA), and Germany's Baader-Meinhof gang. He is suspected of ordering the murder of at least ten opponents of his regime in Europe, the Middle East, and even the United States. Early in 1980 Qaddafi's army, supplied by Wilson's company in Tripoli, successfully invaded Chad, which supposedly has rich uranium deposits.

Federal officials admit that the very latest American equipment and expertise in communications, arms, computer science, and nuclear development is being bought illegally by whoever can pay the price, including Chile, South Korea, Brazil, Argentina, Taiwan, South Africa, Iraq, and Pakistan.

"The full dimensions of the illegal export problem are not even known," said a senior official in the Justice Department. "We frankly don't know how much sensitive technology and military

ARMING THE MIDDLE EAST

Following is a sampling of weapons that were either ordered by or supplied to certain countries in the Middle East between 1979 and 1981. Each recipient nation is followed by a list of those countries from whom it received arms.

EGYPT

U.S.: 40 F-16 fighters; 550 armored personnel carriers; 100 mortar carriers; 12 batteries of surface-to-air missiles; 1,520 antiaircraft missiles; 244 medium tanks. *Britain:* 18 Hovercraft. *China:* 30 F-7 fighters.

ISRAEL

U.S.: 75 F-16 fighters; 250 medium tanks; 100 surface-to-air missiles; 600 air-to-surface missiles; 600 antiaircraft missiles; 600 surface-to-surface missiles; 30 helicopters; 800 personnel carriers; 5,000 antitank guided weapons.

IRAQ

France: 24 Mirage fighters; unknown number of antiaircraft missiles and antitank guided weapons; 46 Puma helicopters. *Italy:* 4 Lupon-class frigates; 6 corvettes; unknown number of Vesuvio-class support vessels. *Switzerland:* 40 training aircraft. *Soviet Union:* unknown number of MiG-23, MiG-25 and MiG-27 fighters. *Yugoslavia:* 1 frigate.

JORDAN

Britain: 279 medium tanks. *U.S.:* 6 F-5 training aircraft; 78 155-mm and 29 203-mm howitzers; 78 armored personnel carriers.

equipment is being smuggled to foreign countries. The dimensions of the problem are frightening."

So far no one's been able to do much to stop arms and technology smuggling because of bureaucratic delays and interagency feuding. Congress is supposed to monitor any arms sale that totals $25 million or more, as well as sales amounting to $7 million to countries outside NATO. But there's nothing to stop a manufacturer or arms broker from breaking a big order into smaller shipments that would escape review.

Military personnel in American embassies abroad are charged with seeing to it that the arms we do sell arrive at their proper destination and remain in the hands of the government

LEBANON
Italy: 6 helicopters; 6 fast attack craft. *Rumania:* 400 rocket launchers; 10,000 rockets.

LIBYA
Canada: 10 light aircraft. *France:* unknown number of antiaircraft missiles. *Spain:* 3 submarines.

SAUDI ARABIA
France: unknown number of surface-to-surface missiles and surface-to-air missiles; 200 infantry fighting vehicles. *Spain:* 20 transport aircraft. *U.S.:* 62 F-15 aircraft; 32 tanks; 1,170 antitank guided weapons and over 1,000 launchers; 916 air-to-surface missiles; 660 antiaircraft missiles.

SYRIA
Soviet Union: 250 T-72 tanks; unknown number of MiG-27 fighters and surface-to-surface missiles.

YEMEN
Poland: 200 medium tanks. *Soviet Union:* 300 medium tanks; 40 MiG-21 fighters; 20 Su-22 fighters; 150 armored personnel carriers; more than 65 rocket launchers.

(*Note:* This information, which was reported in *The New York Times,* came from the International Institute for Strategic Studies.)

that bought them. But foreign governments frequently buy arms and then resell them without telling the original seller. Iran sold American patrol boats to the Sudan; Israel sold jets with American engines to Honduras; and Libya sold American jets to Turkey—all without American knowledge or consent. It is all too clear that when the United States sells arms, it frequently has no way of knowing who will end up with them and how they will be used.

Even with the tightest control of arms sales we still face the problem of theft. A British study group recently reported that the United States is the world's largest supplier of arms to terrorists—through thefts from its armories. Guns stolen from U.S.

Army depots in Europe have ended up in Northern Ireland, Bangladesh, and El Salvador.

Between 1971 and 1974 enough arms disappeared from U.S. arsenals to equip 8,000 men.

By blessing unstable regimes with sales of our sophisticated technology, we end up canceling the technological edge we work so hard to attain. In 1974, for example, the Shah of Iran bought from us some F-14 jet fighters equipped with highly classified radar, which the Iranians claim to have removed and safeguarded. But because we now feel sure that they have developed countermeasures against our radar, we have been forced to make changes in the radar circuitry on our own aircraft as well as in the circuitry of the Phoenix missiles that the radar guides. The changes cost American taxpayers over $285 million.

As Third World countries acquire more expertise, some, including Brazil, Argentina, South Africa, Israel, and India have themselves begun to export advanced weapons technology. Some of these countries built up their own defense industries from coproduction arrangements with the United States. Now they're in a position to compete with us, and to make their own contributions to regional instability. Israel sells everything from machine guns to missiles, to the tune of $500 million to $1.3 billion a year, or 2 percent of all arm deals worldwide. Although Israel won't name its clients, they supposedly include South Africa and Argentina, both fiercely anti-Zionist regimes. Israel also sold arms to Iran for the war against Iraq that began in September 1980. When criticized for their arms sales policies, the Israelis reply that they are no more cynical or money hungry than the superpowers.

Israelis claim they need to sell arms in order to beef up their economy. Economists warn, however, that when a country's limited resources of capital and labor are devoted to armaments, little remains for civilian industry and human goods and services—including food. The Council on Economic Priorities, an independent New York research group, is conducting an ongoing study of

ARMS CONFISCATED EN ROUTE TO SOUTH AFRICA

Houston— U.S. Customs agents arrested six foreign nationals May 12 at Houston Intercontinental Airport and impounded a Montana Austria Boeing 707 cargo aircraft bound for South Africa with 2,200 military weapons and a load of ammunition on board.

Arrested and arraigned in a U.S. magistrate's court in Houston for violation of the U.S. neutrality act were four crewmembers of the 707 operated by the Austrian charter airline and two British subjects, said by U.S. Customs officials to be representatives of an international arms trading company. The cargo allegedly was destined for a company in South Africa. Federal officials would not disclose the names of the companies involved in the transaction.

According to U.S. Customs officials in Houston, the arms were purchased from Colt Arms Manufacturing Co., Hartford, Conn., using fraudulent U.S. State Department export licenses. The shipment comprised 636 M-16 automatic rifles, 135 40-mm. grenade launchers, ammunition and caliber .357 and caliber .38 hand guns.

Federal officials, who had been tipped off about the transaction, managed to place agents as drivers of the trailer truck that transported the weapons from Hartford to Houston, and other federal agents also trailed the truck en route.

The suspects were arrested after the truck was wheeled up to the aircraft but before any of the cargo was loaded onto the 707.

The aircraft had been flown from Vienna to N.Y. and then to Houston. Montana Austria is a Vienna-based charter airline operating two Boeing 707-138Bs and one 707-396C cargo aircraft. It also has a subsidiary that is an international travel agency.

Aviation Week of Space Technology

HOW THE U.S. TAXPAYER PAYS

A lesser-known aspect of U.S. arms exports is the untold millions of dollars they cost the American taxpayer through poor management. Except for grant aid, which is funded by taxpayer dollars, the law requires the Pentagon to recover all direct and indirect costs it incurs when it sells weaponry or training through the Foreign Military Sales (FMS) program. But according to the General Accounting Office this is not what has happened.

In 25 reports dating back to 1969, the GAO has consistently found weaknesses in the Pentagon's accounting system that have amounted to subsidies totaling "hundreds of millions of dollars." Foreign governments are not properly billed for training, transportation, or research and development costs. Neither does the Pentagon fully recoup its administrative expenses, inventory losses, or the cost of using government-owned plants and machinery to make arms for foreign customers. The latest in the long series of GAO reports says **the U.S. taxpayer has absorbed "up to an estimated $370 million in costs during the past 6 fiscal years" which should have been borne by foreign governments.**

Source: International Policy Report, Vol. 5 No 1
A publication of the Center for
International Policy
120 Maryland Avenue N.E.
Washington, D.C. 20002

13 major industrialized countries over the last two decades. So far the study shows that countries that spent less on arms have been economically healthier—they grew faster and produced more. The two countries in the study that have put the most into defense, the United States and Great Britain, have the most seriously stagnating economies. Japan, Austria, and Canada had spent less on defense and were in much better economic shape. Ironically, Japan, which has had no defense or arms industry since the end of World War II, finds its peaceful export of cars and cassette recorders threatening to Western economies. "If we had joined the arms race in the Middle East," said one Japanese diplomat, "no one would have complained."

The big American defense contractors swing as much clout in the domestic political arena as they do internationally. Companies like Lockheed, Rockwell, and Grumman have such power in the White House and on Capitol Hill that even a toughie like Office of Management and Budget director David Stockman found it all but impossible to cut defense spending. Arms companies lobby like mad to get Congress to authorize military spending because financially, defense contracts support business ventures in which the corporations experience very little financial risk. The government invests the capital, including funds for research and development, and covers cost overruns as high as 800 percent—a craziness that no other financial backer would put up with. In addition the companies know they can count on the government's continuing demand for new (weapons) products.

"It seems paradoxical that a sector of our economy so privileged as the defense industry could be so wasteful and inefficient," Tom Gervasi notes in *Arsenal of Democracy*, but the reason is fairly obvious. "While the military services, the government, and the taxpayer all experience it as waste, the defense manufacturers do not. *They* experience it as profit."

According to the Government Accounting Office, American defense contractors have averaged 56 percent in profits over the last ten years.

Once the government has authorized a company to begin research and development on a new weapon, it finds it difficult to cancel further development. For their part, arms companies spend flagrantly to keep the fires burning under everyone involved. The eight major defense contractors spent more than $2 million on campaign contributions, between 1976 and 1980, according to "The Iron Triangle," a report from the Council on Economic Priorities. The bulk of the contributions went to friends of the companies on the congressional defense appropriations subcommittees and armed services committees. In addition to making campaign contributions, the big eight (Boeing, Lockheed,

General Dynamics, Grumman, McDonnell Douglas, Northrop, Rockwell International, and United Technologies) also maintain lobbying offices in Washington and even charge some of their costs to the Department of Defense as "administrative expenses." Rockwell spent $7 million in 1974 and 1975 alone on upkeep of its Washington office, and charged the Pentagon $12,500 (among other expenses) for the costs involved in making a ten-minute film on the B-1 bomber—a contract it had won.

According to "The Iron Triangle," two top arms companies, Grumman and Rockwell, conducted grass-roots campaigns in which they urged their employees and their stockholders to write Congress to save weapons programs the companies wanted. In 1977 the Pentagon wanted to cut back the F-14 jet fighter program. Grumman, which had the contract, and which previously had been a relatively discreet lobbyist, abandoned its low profile and launched a hard-hitting ad campaign in the local Long Island newspaper *Newsday* so defense workers would be sure to get the picture. "The F-14 Tomcat. It belongs to you," read the headline. Ad copy went on to describe the proposed cutbacks as a "most serious crisis for the entire Long Island community," because subcontracts would go to about 1,100 Long Island firms plus about 2,000 more in other parts of New York State.

Not surprisingly, in 1978 Congress voted to continue the F-14 project without cutbacks. But Grumman had been so vocifer-

Low-angle side view of a YF-16, armed with one AIM-9 missile on each wing tip in flight.

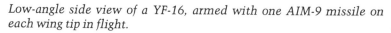

ous in its promotional efforts to keep contracts aloft that Secretary of Defense Harold Brown ordered an audit to see whether any Pentagon funding had ended up in Grumman's campaign chest. (The audit may have been carried out, but the findings were never made public.)

The public relations pays off. Between 1970 and 1979 the eight major defense companies received more than $100 billion in defense contracts.

Defense lobbying is usually effective because legislators are afraid of putting their constituents out of work. This is especially true of senators and representatives from California, the arms capital of the United States (home of Lockheed, Northrop, and the electronics companies in Silicon Valley), the South (site of a number of defense plants), and the state of Washington (Henry Jackson has been called "the Senator from Boeing"). Critics of defense spending maintain that defense contractors don't really support such large labor forces, tending to keep labor costs down so that profits stay high. Tom Gervasi says, "The more specialized an industry becomes, the more capital intensive. The defense industry has a decreasing requirement for labor."

British journalist Anthony Sampson, who wrote *The Arms Bazaar*, points out that arms companies and their agents usually defend themselves with the claim that they're only "doing the necessary" in making arms, and it's a shame human nature is such that people actually use arms to kill one another. But jumps in arms production and sales, according to Sampson, are usually dictated by economics, whether domestic or international. "The scale of the arms race could have been very different if Nixon had said no to the Shah." Governments often cover up the full extent of the arms business, but the public needs to be told what's going on. As Sampson observes, "the arms trade, like narcotics or slavery, is different from other trades." Its problems are moral as well as political.

Stored plutonium at Rocky Flats, a plutonium production plant in Denver, Colorado. Before 1940 there was no plutonium on earth. Now there are hundreds and hundreds of tons of it.

chapter 13

Proliferation: The Spread of Nuclear Power

On June 7, 1981, Israel bombed and destroyed an Iraqi research reactor. Since that day President Reagan has had to rethink the problem of proliferation, an issue he told voters during his campaign was "none of our business." The entire world was riveted by Israel's rash behavior. Convinced that Iraq, an avowed enemy for more than 30 years, would produce nuclear bombs from its new reactor, Israel panicked, and the complex and troubling issue of proliferation lifted once again to the forefront of international consciousness.

Iraq had acquired its peaceful nuclear technology legally and openly, from France and Italy. A country rich in oil, Iraq had no domestic nuclear program and, indeed, no need for one. Its precise plans for its reactor were never stated. Most American experts doubt that the destroyed installation could ever have produced bombs on an efficient, ongoing basis, but believe nevertheless that it could have managed to produce a few within some years' time. And how many bombs does a country need?

However well grounded Israel's fears may have been, its "nuclear Entebbe" was at the very least a vote of no-confidence in the international safeguards that are meant to confine nuclear technology to peaceful use. Ironically, unlike Israel, Iraq had agreed to respect international nuclear controls. By current standards Iraq has been a model nuclear citizen. In 1968 it signed the

Nuclear Nonproliferation Treaty, swearing not to develop nuclear weapons from peaceful technology, and opening its installations to inspectors from the Vienna-based International Atomic Energy Agency (IAEA), which enforces the international safeguards. (For the most part these "safeguards" are little more than accounting records of nuclear fuel, backed up with plant inspections and monitoring devices.) Iraq's signing of the treaty notwithstanding, Israel suspected its enemy was almost certainly bent on building a bomb.

Before the Israeli raid the Reagan administration had been reviewing President Carter's relatively stiff antiproliferation policy. The objective had been to ease some of a 1978 law's restrictions on exports of nuclear technology and fuel to "reliable" developing nations pursuing peaceful nuclear programs. Since the bombing of Iraq's reactor, however, Reagan has been feeling increasing pressure to tighten restrictions and step up international efforts to stem the flow of nuclear fuel and sophisticated equipment.

Ever since the destruction of Hiroshima and Nagasaki there has been universal agreement that nuclear weapons must never again be used in an attempt to settle disputes between nations. At the same time, virtually all nations agree that because gas and oil resources are limited, nuclear power is the most promising source of energy in the coming century, that its development is highly dangerous, that it will be developed nevertheless for peaceful purposes and in the process will make production of nuclear weapons possible. In other words, the idea of pulling proliferation to a fast halt has become obsolete. The most we can hope for is finding ways to cope in a partially proliferated world.

Every American administration since Truman's has wrestled with this dilemma. Years of post–World War II negotiations ultimately produced the Nuclear Nonproliferation Treaty of 1968. The treaty did create a framework for inspecting nuclear facilities, but its flaws include the fact that even signatories are permitted to develop weapons under the threshold of the treaty's restrictions. What's more, any signer can denounce the treaty on 90 days' notice if it feels national security so demands. At present the treaty has 112 signatories, but it has not been signed by some of the most important potential-nuclear and almost-nuclear powers in the world, including Israel, Argentina, Pakistan, Brazil, South Africa, China, India, and France.

Although the spread of sensitive nuclear technology has slowed since 1968, it will probably never be brought to a complete halt. There are too many reasons why critical nations won't sign the treaty. First, regional conflicts (in southern Africa, the Middle East, and southern Asia, for example) give many nations the incentive to own, if not to use, nuclear arms. Second, nuclear Have-nots have noticed that the superpowers sometimes renege on guarantees of aid when smaller allies find themselves beleaguered. For this reason China, France, India, Israel, South Korea, and Iran have each moved to develop first-strike capability. (Japan, which currently has no defense program at all, may eventually join them.) Last, and far from least, many nuclear Have-nots want to be able to deal with the superpowers on an equal footing. A nuclear arsenal is a sign of a nation to be reckoned with. Today six countries admit to possessing nuclear weapons, the United States, the Soviet Union, Great Britain, France, China, and India. By 1987, according to the Energy Research and Development Administration, 30 more will have the materials and the technology to make such bombs.

> The major promoters of nuclear power have contracts at DOD facilities for the production of nuclear weapons. These corporations include General Electric, Westinghouse, Dupont, Union Carbide, and Rockwell International. For example, Rockwell receives a nominal management fee after costs for managing the Rocky Flats plant near Denver, Colorado. Rocky Flats is a plutonium fabrication facility that also engages in federally subsidized research. What makes facilities like Rocky Flats so attractive to their corporate operators is the opportunity to gain—at no cost to themselves—a vital experience base for their commercial nuclear reactor programs.
>
> Pam Solo and Mike Jendrzejcyk
> *Business and Society Review*
> Spring 1980, No. 33

Between 1970 and 1975 the United States has more than doubled the number of weapons in its nuclear arsenal. At present it has about 31,000 nuclear warheads and is manufacturing new ones at the rate of 3 a day. From 1945 through 1980 there have been 1,274 nuclear explosions, 667 carried out by the United States, 447 by the Soviet Union, and the remaining 8 percent by France, England, China, and India. The majority of these test explosions have taken place *since* the partial test-ban treaty of 1963, and they are continuing at the rate of between 34 and 42 explosions per year. At this rate it is estimated, there will easily be 10,000 new nuclear weapons stashed in various countries within the next decade.

The stockpiling of further nuclear weapons by countries that already own some is called *vertical* proliferation. The spread of nuclear weapons to new countries is referred to as *horizontal* proliferation. Ironically, horizontal proliferation got a firm foothold in 1953 with "Atoms for Peace," President Eisenhower's attempt to beat swords into plowshares. His program was aimed at solving the world's energy problem by making American nuclear know-how available so that other countries might build their own reactors. Unfortunately "Atoms for Peace" ultimately led to a dangerous proliferation of all forms of nuclear power. As has been learned retrospectively, once a country has a reactor, the steps to building a bomb factory are few and short. The overlap between peaceful and military uses of nuclear power is virtually inevitable.

* Domestic nuclear power requires knowledge, skill, and equipment that are in large part the same as those needed for making bombs.
* Uranium needed for atomic fission is no longer hard to come by. Either natural or low-enriched uranium can be enriched to the 90 percent-plus concentrations needed for making bombs, and new laser technology will make the enrichment process considerably easier in the near future.
* Fissionable materials—notably plutonium—needed to make bombs can be extracted quite simply from reactors' spent fuel. British physicist Amory B. Lovins, a vigorous proponent of energy alternatives to nuclear power, notes

that domestic nuclear power all too easily provides "an innocent civilian cover" for bomb making.

More and more nations have acquired the technical knowhow and fissionable material from which bombs can be made. Today at least 15 countries, including politically unstable ones such as Iran, Egypt, Argentina, and South Korea, are capable of producing bombs within a short period of time should they so choose. American political experts think that Iraq, Pakistan, Libya, South Africa, South Korea, and Taiwan are trying hard to make nuclear weapons, either through secret purchases of materials or through indigenous technological ingenuity. In spite of strenuous denials both Israel and South Africa are suspected of actually having made nuclear weapons. (At the moment, Brazil's and Argentina's nuclear programs appear to be aimed at commercial energy needs.)

Proliferation is due for a sharp increase as a result of the recent development of a new generation of reactors called "breeders," which actually end up producing plutonium. Besides being extremely poisonous, plutonium can be used in weapons in its natural state, requiring no "enrichment" or other process to make it bombworthy. In countries with breeder reactors the process of diverting the fuel for weapons construction could be so rapid that there would be no time for diplomatic or other protective measures should these countries feel compelled to go quickly to war.

There are further hazards inherent in the spread of nuclear technology. Any country with nuclear power plants or fuel facilities runs the danger of sabotage at poorly guarded sites. Recipes for a crude atomic bomb are now available in technical journals, encyclopedias, and other public literature. In 1976 John Aristotle Phillips, a Princeton undergraduate, designed an atomic bomb as part of a college physics research project. In 1978, free-lance journalist Howard Morland published "the H-bomb secret" in *The Progressive.* Both used research materials available to anyone. Any nation, terrorist, criminal, or lunatic need only to acquire the requisite materials for building a nuclear bomb; the information on *how* to do it is plainly available.

A crude terrorist version of World War II's atomic bomb

would pack only about one thousandth of the punch of the modern hydrogen bomb, but it could be delivered to an American city in the trunk of a car. If parked near a densely populated area like New York's financial district, for example, and exploded during office hours, a homemade nuke would kill as many as a million people. Detonated in a more central location on Manhattan Island, it could throttle the city in all directions for 3 miles. A terrorist atomic bomb would be likely to be exploded near or on the ground, generating vast amounts of fallout and rendering the city uninhabitable for months or years—perhaps indefinitely. (By contrast, the Hiroshima and Nagasaki bombs were airbursts that kept fallout negligible and permitted reoccupation of the cities within days.)

Until late in the last decade, the weakness of international safeguards against proliferation was known only to a select group of nuclear officials, diplomats, and intelligence agents. But the truth has begun to go public as a result of some particularly dramatic episodes.

In 1977 the press finally learned about the "Plumbat affair" which had occured nine years earlier, in 1968. A merchant ship en route from West Germany to Italy disappeared from the high seas. **When it reappeared several days later, the ship was bearing a new name, flag, captain, and crew, and its cargo of 200 tons of uranium, in drums mislabeled "Plumbat" (instead of Plumbum—lead), had disappeared.** This kind of nuclear skullduggery can take place because there is no international escort system that would require accompanying vessels carrying low-grade or bomb-grade nuclear materials. Although the uranium lost in the Plumbat affair was never officially traced, the intelligence community finally concluded that all 200 tons had gone to Israel. Even if the stuff were used for peaceful means—say, in research reactors operating in Israel and India at the time—200 tons of uranium would have produced enough by-product plutonium to supply as many as 40 atomic bombs.

The Israeli government has stated repeatedly that it "will not be the first country to introduce nuclear weapons into the Middle East." But the CIA suspects—though so far it has made no official announcement—that Israel has been making nuclear weapons since 1974. Under the present international system

Where will we be in a decade if the nuclear arms race continues?

* 75,000 nuclear warheads—50 percent more than today—in the U.S. and Soviet arsenals, with an explosive power of over 2 million Hiroshima bombs.
* more than 10,000 missiles and bombers, in the U.S. and Soviet nuclear forces—double the number today.
* over $300 billion spent on nuclear missiles, aircraft, and bombs by the United States alone—$6,000 from each American family.
* small, concealable weapons systems that could deliver thermonuclear warheads unverifiable by agreements or ordinary detection systems. Today, bombs with Hiroshima-level destructive power can fit in a 6-inch shell or rocket. They use only a grapefruit-size amount of plutonium.
* ten or more nations—some highly volatile—may have nuclear weapons. Terrorist groups may have acquired Hiroshima-size bombs.

there is virtually no way to account for missing nuclear materials. At the end of 1976 a total of 8,000 pounds of highly enriched uranium and plutonium were missing from U.S. nuclear facilities. Theoretically the missing material was enough to make hundreds of clandestine atomic weapons. Missing uranium is called MUF (Materials Unaccounted For) in the jargon of the Nuclear Regulatory Commission (NRC), the federal agency that oversees the domestic nuclear industry and has independent power to license nuclear exports. One of the most disturbing MUFs was a 1965 loss of 200 pounds of highly enriched uranium from a nuclear facility in Pennsylvania. Again, the theft was deduced to be the work of Israeli intelligence.

In its very attempt to prevent nuclear power from ever again being applied to military purpose, the United States ended up becoming the Johnny Appleseed of nuclear technology. It was clear that the United States couldn't prevent nations from developing nuclear power on their own. The hope was that we might at least steer them in a peaceful direction. From Eisenhower's announcement in 1953, through late 1974, the United States supplied grants, funding, and technology for the building of reactors and trained thousands of foreign nationals in nuclear physics. Of course, this "peaceful" program was also a great financial boon to the American nuclear industry. Suppliers like Westinghouse and General Electric still receive most of their reactor orders from abroad.

The "Atoms for Peace" campaign ended abruptly in 1974 when a coded message, "The Buddha is smiling," was flashed to New Delhi from a remote atomic testing site in the desert of western India. Exploding a 15,000-kiloton nuclear test device—slightly larger than the Hiroshima bomb—India became the sixth member of the nuclear superpower club. At first India, which had refused to sign the Nonproliferation Treaty, claimed that the device had been made with Indian materials only. Two years later it became known that India's bomb had been made with plutonium produced by a Canadian-supplied research reactor and "heavy" water provided by the United States.

India was open about the fact that it craved the power and international prestige associated with owning nuclear weapons. One Indian official had this to say about the situation of the nuclear Have-not: "If we make the bomb, the United States will realize that they cannot ignore a nation of seven hundred million people, just as Nixon said about China in 1971. It will enable us to deal with China on an equal basis. We shall be able to sustain our cordial relations with the Soviet Union, too."

Largely in reaction to India's 1974 nuclear explosion, President Carter made proliferation slowdown a top priority of his administration. He halted the domestic development and export of breeder technology, and in 1978 he signed the Nuclear Nonproliferation Act, which Congress had passed almost unanimously. Among other provisions the 1978 law sought to limit U.S. export of highly enriched uranium. The NRC could refuse to grant a license for such export; the agency could be overruled by the President, but he in turn could be vetoed by Congress. Through the law, Carter pressed for the development of an alternative to highly enriched uranium and tried to halt the stockpiling of plutonium and the spread of reprocessing technology. He also worked to dissuade nuclear exporters from providing equipment or training to nations that would not forswear the development of nuclear weapons.

While Carter's nonproliferation policy scored some successes, the 1978 law was difficult to enforce.

* The United States, which exports low-enriched uranium to countries that agree not to enrich it to bomb levels, cannot control the enrichment business around the world. West Germany has sold enrichment technology to Brazil; France has sold highly enriched uranium to Iraq; and Brazil is suspected of clandestinely shipping uranium to Iraq. Britain, Japan, France, and West Germany are all working on plutonium reprocessing.
* The law has no way of covering all the various methods of bomb making that might be conceived in the future.
* Other nations resent the law's implication that some countries are more trustworthy than others. And. . .

Someone is always ready to argue for a worthy exception to the rules.

In fact, just over two years after the law was enacted, Carter himself made an exception to the rules when he overrode the NRC's veto and congressional objections and okayed the shipment of 38 tons of uranium to India. The President argued that this move was a calculated gamble to win India's good will and buy time for diplomacy that might bring its weapons under better control. His critics fear that he set a precedent for future exceptions for Brazil, South Africa, and Spain, which, like India, have refused to agree to countrywide reactor inspections. Today India is visibly preparing for a second bomb test.

When in late 1980 the Soviet army occupied Afghanistan, Washington's priorities shifted overnight. India's archrival, Pakistan, right next to Afghanistan, was in a difficult geographic position and wanted U.S. aid to ward off Soviet invasion. At the same time, Pakistan had made no bones about wanting a bomb because India has one. The Pakistanis have already openly purchased sensitive technologies and are now building a tunnel in which to test a nuclear weapon.

The thought of Pakistan's leader, General Mohammad Zia ul-Haq, a volatile and particularly repressive chief of state, owning a nuclear weapon is unsettling. President Reagan has nonetheless offered Pakistan a $500 million package of military and economic aid over five years. In order to help Pakistan, he must, like Carter, persuade Congress to modify the terms of the 1978 law forbidding assistance to countries that seem to be developing nuclear weapons. Reagan wants major changes in the law, especially the elimination of the NRC's independent export license.

Reagan's opponents believe that whatever military aid goes to Pakistan should be predicated on a halt to its nuclear weapons program. Recently Senator John Glenn, Democrat of Ohio and sponsor of the 1978 act, defended the law and emphasized that proliferation inevitably makes the prospect of nuclear war more likely:

> *Right now there are 247 nuclear power reactors in the world. Each develops about 146 gigawatts (1 gigawatt equals 1 billion watts) of power. Each of these gigawatts develops about 250 kilograms of plutonium each year. A nation with fairly low sophistication in nuclear matters can make a bomb from 5 kilograms. That means you could have between 25 and 50 bombs made from the leftover fuel of each gigawatt each year. If you multiply all that, the potential from 247 nuclear reactors is between 4,000 and 8,000 bombs per year. That's what we're trying to prevent.*

Under Carter, America's policy on proliferation was schizophrenic and disingenuous at best, but with Reagan in office the tone has become alarmingly cynical. It's possible that administration officials talk of waging "limited" nuclear war to justify what is simply inevitable if proliferation continues unchecked. Since the Israeli strike against Iraq, Reagan's officials have been saying that their new policy would endorse the reprocessing of plutonium in "reliable" countries. They recommend specifically that Japan and Western European allies be permitted to reprocess spent fuel to ease their need for imported oil. The new policy would also call for tight controls on the sale of sensitive technologies to suspect countries, but would be more "discriminating." Some nonproliferation advocates fear this means that the Reagan policy would not be restrictive enough, and that the administration will view nonproliferation as another form of arms control, for which it has shown little enthusiasm.

One nonproliferation advocate, Earl C. Ravenal of Washington's left-wing think tank, the Institute for Policy Studies, says, "The crux of the matter of nonproliferation is enforcement." As they stand, the nuclear safeguards administered by two international agencies, the IAEA and Euratom, are scandalously weak and full of holes.

The problem of proliferation points up both the responsibilities and the fragility of the global village. Says Ravenal: "No one can prescribe completely satisfactory controls on nuclear weapons and nuclear materials—or avoid completely the task of devising them. No nation can confidently implicate itself in international political regimes to determine other nations' acquisition and use of nuclear arms and materials—or completely escape from the consequences of other nations' malfeasance." Ravenal emphasizes that the United States must set an example by restraining its own conduct but should avoid "either forcible measures to restrain other countries or the extension of nuclear guarantees and other military commitments to secure their compliance in a regime of nonproliferation."

Ultimately, nonproliferation may depend less on controlling technology than on refusing regional conflicts. Senator Daniel P. Moynihan of New York has observed that while the challenge of the 1970s was to stop nations from obtaining nuclear weapons, the 1980s will need to find ways to stop countries from *using* them.

An Air Force B-52 sits atop TRESTLE, a facility that can electronically simulate the electromagnetic effects of a nuclear explosion on an aircraft and its electrical components. TRESTLE will simulate this energy by using two 5-million-volt pulsers that will discharge voltage into wire transmission lines surrounding the aircraft. Sensors will capture aircraft EMP response signals, and these signals will be transmitted to computers inside a shielded enclosure.

TRESTLE is built from 1-foot-by-1-foot wooden columns connected by wood crossmembers and held together with approximately 250,000 wooden bolts. It is believed to be the largest glued-laminated wood structure in the world. Over 6 million board feet of lumber were used—enough to build 4,000 frame houses.

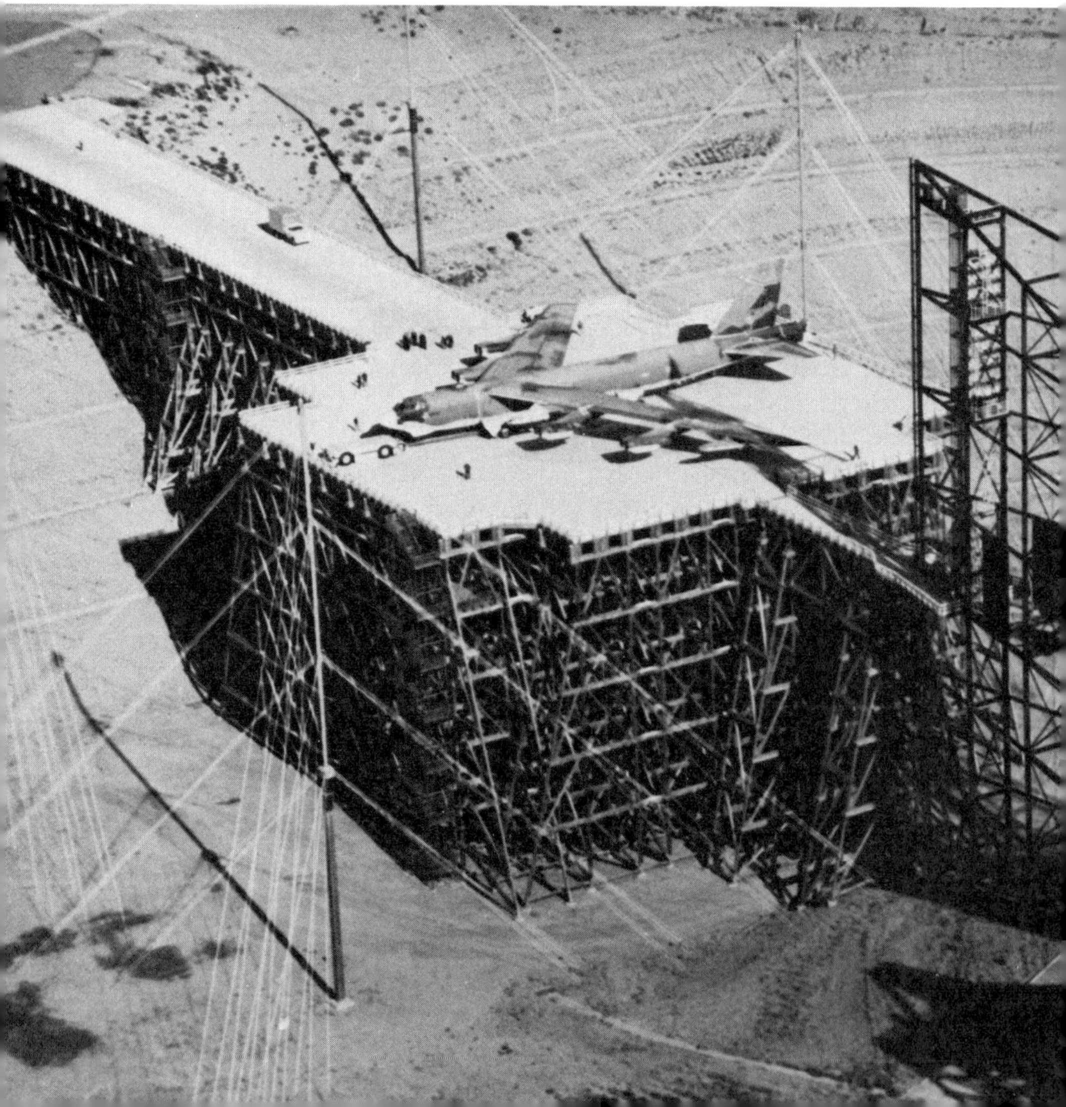

chapter 14

Nuclear Chaos and the Threat of EMP

In July 1962 the United States tested a 1.4-megaton H-bomb high above the atmosphere, some 258 miles over Johnston atoll in the Pacific. In Hawaii, 800 miles away, something bizarre (and at that time inexplicable) happened simultaneously. Streetlights went out, burglar alarms went off, and circuit breakers in power lines popped open.

Earlier, in 1958, a forewarning of this episode occurred when nuclear detonations were set off closer, at 27 and 48 miles over the same island. Disturbances in the ionosphere seemed to affect radio communications and radar. At this time the U.S. military had planned to conduct nuclear tests higher up to see if an even worse blackout would occur, but a test-ban treaty struck the same year forestalled the research needed to find out what these disturbances were all about and how devastating their effects might be. When tests were briefly resumed in the early 1960s (a partial ban went into effect again in 1963), they indicated that the 258-mile test over the Johnston atoll had indeed been related to the peculiar events in distant Hawaii.

Today the mysterious agent that creates these disturbances is known as EMP, or electromagnetic pulse. In 1963 physicists at the Rand Corporation first reported on EMP and described it as the phenomenon that accounts for the strange radio and radar blackouts that occur following high-up nuclear explosions. Discussed recently in *Science* magazine, which ran a long series warning of the Pentagon's laxity in dealing with the problem, the scientific explanation of EMP was given as follows:

Earthbound gamma rays from a nuclear explosion in space eventually hit air in the upper atmosphere and knock out Compton electrons, which are deflected by the earth's magnetic field and forced to undergo a turning motion about the field lines. By a complex mechanism, these electrons emit EMP, which at ground level can radiate over thousands of miles with a peak strength of 50,000 bolts per meter. Any metal object picks up the pulse. If the object, such as an antenna or a cable, leads to sensitive electronic components, the pulse can cause extensive damage.

Any nuclear explosion produces some EMP, scientists found, but only bursts that occur outside the earth's atmosphere are capable of producing microsecond pulses of far-ranging damage on the earth below. In fact . . .

One of these split-second pulses is 100 times stronger than a lightning bolt.

As might be imagined, the implications of this discovery were highly disturbing to the American military. If, in battle, the Soviets should strike first—detonating a nuclear warhead 200 miles above SAC headquarters in Nebraska, say—the resulting EMP would knock out unprotected communications equipment across the entire United States. Although the President is supposed to have other communications channels at his disposal (43, in sum), all of these might conceivably be affected by EMP from a Soviet blast.

For some time military scientists have been privy to information on which the Pentagon has been notably slow to act: If the USSR wanted to destroy our entire communications network, it wouldn't even have to launch a missile. The picture-taking Cosmos satellite could do the job. Routinely crisscrossing the United States, in orbits of 200 to 450 kilometers above the earth, the Cosmos, should it be fitted out with a couple of pounds of plutonium and detonated, would produce enough EMP to damage our total power grid.

America would lose all electrical equipment without its own power supply. Traffic lights, computers, radio and TV. Telephones and military channels would shut down.

In short, electromagnetic pulse makes a mockery of the 11 to 15 minutes the President is supposed to have for ordering a counterattack. The chance definitely exists that he would have no way of issuing a call to arms.

Precisely how damaging EMP could be is likely to remain a mystery until adequate tests are conducted. Since the resumption of the partial test-ban treaty in 1963, most information about EMP has had to be acquired through simulations. Testing real nuclear explosions in the exoatmosphere would tell us a lot more than we know now. The United States is ready to begin these tests at a moment's notice. The Defense Nuclear Agency spends $11 million a year to keep a small outpost on Johnston atoll ready and waiting for real tests. Should the current ban be rolled back, a 165-member task force would immediately begin test-firing nuclear-tipped missiles into space and, among other tests, measuring the amount of EMP these detonations produce.

The Pentagon is worried that the Russians may know more than we do about EMP. *Their* high-altitude tests were carried out over central Asia, which, though sparsely populated, does have some cities, and these would have been sufficient to allow Russian physicists to study the effects of EMP on earth-based electronics systems. As far back as 1968, DOD was alarmed to discover, Russians were onto EMP, as evidenced by a Soviet Ministry of Defense publication reporting that "powerful nuclear explosions set off at great altitudes" constitute a considerable threat to ICBMs "because the impulses of electromagnetic energy created by such explosions can put out of commission not only the on-board missile equipment, but also the ground electronic equipment of the launch complexes."

There is reason to believe that the Soviets may already have protected their weapons against EMP. In 1976 a Soviet defector flew his MiG-25 to Japan. Later, upon investigating the plane, the Pentagon noticed that the aircraft's computer circuitry was technologically up-to-date except for the fact that it used old-fashioned computer vacuum tubes, which are far less sensitive to the effects of EMP. Thus the Pentagon concluded that the Soviets may have known how to "harden" their planes against EMP. (By the 1970s physicists had discovered that the old computer vacuum tubes are about *10 million times more resistant* to EMP than are modern semiconductors.)

Today our government is pouring money into the "hardening" of electronic instruments against EMP by means of steel shielding, special grounding and circuit layout, and perhaps even by reverting to the use of vacuum tubes. All of this will end up costing billions of dollars. And there is still educating on EMP to do in the Defense Department itself. Gordon Spear, scientific advisor to the Defense Nuclear Agency, which has been trying to research the EMP problem, says his agency's work is complicated by a cynical philosophy "that says nuclear war is never going to happen. This leads people to say that only Soviet *perceptions* of U.S. capability are important. ABM is important. EMP hardening, on the other hand, is not very impressive, and declaring that a vulnerable system is hard is probably as effective as hardening the blasted thing."

Some scientists believe the Pentagon and its contractors know that they're building self-defeating systems but justify what they do on the grounds that the weapons will never have to be used.

THE POSSIBILITIES FOR PROTECTION AGAINST EMP

Officially the Pentagon claims that the Emergency Action Message system, which the President would use to deliver the doomsday signal, is "especially designed to endure the effects of jamming, physical destruction, nuclear blackout, and electromagnetic pulse." Yet the giant ground-based military communications system uses a great deal of metal in its copper cables, microwave towers, switching centers, and command posts, and metal is what attracts EMP. In addition, the vastness of this system makes it virtually impossible to test all of its various parts and subsystems for hardness against EMP. The few such tests that have been conducted so far are not very encouraging. For example, Autovon, a military network that was built for the government by Bell Labs, has been described in one of the major communications textbooks as being "nuclear bombproof." But in 1975 an Autovon switching center was exposed to simulated bursts of EMP and suffered considerable damage as a result. James Kerr, a government specialist in EMP, says, "We know how to protect to a certain extent, but the EMP problem is not susceptible to cure."

More reliable than ground-based communications systems are those based on satellites. The Pentagon relies on satellites for more than 70 percent of its long-distance communications and it is planning on more. For example, the 1,000 Minuteman missiles scattered around the country in silos will soon have their launch control centers equipped with satellite ground stations. And at present 400 Navy ships have satellite links, as do almost all B-52 bombers. The airborne command posts, or AWACS, "looking glass" planes of the Strategic Air Command that constantly circle over the United States ready to order a nuclear first strike or counterattack, have satellite capabilities for doing so. But while they are less likely to be put out of commission by EMP, satellite communications have other problems. They're vulnerable to Soviet "killer satellites."

Also, satellites haven't been sufficiently tested against the effects of EMP for the Defense Department to consider them utterly reliable. The first full-scale test took place only recently, in

1980, at the underground test site in Nevada. As yet it is not known whether satellite signals are able to penetrate ionospheric disruptions caused by high-altitude nuclear explosions.

What can the government do about the threat of EMP? Our ground-based communications network may never be hard enough; satellites can be blasted out of the sky by other "killer satellites"; the government's *only* contractor for defense communications, Bell Labs, lags behind smaller companies in developing EMP-resistant technologies. Meanwhile, the communications gap with the Soviets is rapidly becoming a political football, the 1980s version of the "missile gap" of the 1960s. Doves claim that closing the communications gap (including hardening our networks against EMP) is impossible. Hawks claim that the gap can be closed, and that this must be done as soon as possible—regardless of cost—for the sake of national security.

Some scientists, among them Edward Teller, inventor of the H-bomb, would like the government to get the 1963 test ban waived so that more could be learned about EMP. Doves use the existence of EMP as further evidence that the new strategy of "limited nuclear war" is completely unfeasible. The communications chaos caused by the first bolts of EMP would mean that we could not participate in a nuclear back-and-forth but would have to fire off our entire strategic arsenal at the start. In other words, only a "use it or lose it" philosophy would apply. EMP would swiftly eliminate the precise, elaborate command systems needed to observe the effects of nuclear strike and direct a limited war, making George Bush's idea that "you can have a winner" in a nuclear exchange a dangerous pipe dream. As John D. Steinbrunner, a senior researcher at the Brookings Institution put it:

"The precariousness of command channels probably means that nuclear war would be uncontrollable, as a practical matter, shortly after the first tens of weapons are launched."

It is difficult for anyone involved to know what, at this point, the Pentagon should do about the complicated threat of EMP. Critical questions continue to go unanswered. One is whether satellite signals would actually be able to get through the ionospheric disturbances caused by high-altitude blasts. Another has to do with ground links. According to Harry Griffith, head of the Defense Nuclear Agency, "Full testing of the communications-link disturbances caused by high-altitude nuclear weapons are not even scheduled until 1982. Actual hardening will take longer." In the meantime the President's airborne command posts, four especially designed Boeing 747s, are notably weak as a command channel since only one of them is hardened. According to the *Science* report on EMP:

"The other three, on call 15 days out of every month, have as many as 11,500 essential circuits that would fail if the planes were hit by an electric pulse from a nuclear burst thousands of miles away."

The Pentagon says dozens of alternative methods exist by which the President can communicate with nuclear forces, including cables, satellites, microwave relays, and certain types of radio transmitters. But continuing testimony in Congress makes clear that all is not well. Gerald Dinneen, who in 1980 was the Pentagon's top specialist on defense communications, suggested that public knowledge of the vulnerability of our communications system should be squelched. "I think discussions of these things should be held in closed sessions," he said. "Some of the comments about the weaknesses of our command and control system must be kept at a very high level of classification."

Transportation of nuclear waste through population centers has sparked a controversy over the risks of accidental spillage.

chapter 15

Broken Arrows, Bent Spears, Dull Swords: The Truth About Nuclear Accidents

Accidents involving nuclear weapons have occurred with disturbing frequency since at least 1950. The Pentagon classifies such accidents according to degree of danger. The least dangerous is a Dull Sword, or a "nuclear weapon minor incident." A "nuclear weapon significant incident" is a Bent Spear, a label for heart-stoppers such as the day in June 1974 when a helicopter carrying nuclear weapons had to make an emergency landing near New York's popular Jones Beach. An out-and-out nuclear weapon *accident* is a Broken Arrow (a phrase American Indians used to indicate peace).

Mercifully none of the accidents so far have been in the worst possible category, a NUCFLASH (pronounced "nuke flash") or "nuclear weapon war risk accident"—an accident, that is, that another country could mistake for an attack. Nor have any Broken Arrows involved an actual nuclear detonation. The Pentagon doesn't rule out that possibility, but it insists that the chances are remote.

The risk of a plutonium leak from a damaged bomb, on the other hand, is not remote at all. Several Broken Arrows have spread radioactive materials over wide areas. What happens is

NUCLEAR CONVOY

Ordinarily, the convoys glide by unnoticed on the freeways, the unmarked trailer trucks blending into the traffic flow and their escort vans looking much like campers off for a weekend jaunt. The drivers listen to the CB radio, put in at the weigh stations, observe the speed limit, and pay the occasional traffic ticket. They pull up at the truck stops for meals and gas.

But these drivers are no ordinary breed of highway haulers. They come equipped with pistols, rifles, shotguns, machine guns, and hand grenades—and they are trained to use them. They are linked by high-frequency radio to each other and to a national command post in Albuquerque which monitors their movements around the clock.

Their vehicles, designed and tested at Sandia labs, are mobile vaults. They are built to withstand head-on collisions at sixty miles per hour without their cargoes so much as shifting, and to endure an 1,850 degree fire for half an hour (the equivalent of a crash with a gasoline tanker) without raising the temperature inside. Their metal linings are designed to resist drills, blow torches, and explosive charges. Should they be penetrated, an assortment of foams, sprays, and other dirty tricks awaits the intruder.

"No one has ever tried, and we hope they never will," said Donald P. Dickason, the man in charge of the Energy Department's massive fortified transportation system for nuclear weapons and the parts that go into them.

Samuel H. Day, Jr.
The Progressive
Research assisted by a grant from
the Fund for Investigative Journalism

that the TNT in the A- or H-bomb goes off, killing and injuring people, and cracking the bomb so that uranium or plutonium spills into the sea or becomes scattered over the countryside.

Plutonium's radioactive half-life is about 24,000 years. It is so deadly, a person who inhales one millionth of a gram is likely to get lung cancer.

There are at least 4.5 pounds of plutonium in most atomic weapons.

Most of the 35,000 nuclear weapons in the United States are scattered throughout 40 states. Often they are stored in or transported through heavily populated areas. Accidents are most likely to occur during transport. The United States moves hundreds of weapons every day. Yet it takes few precautions to protect even the people who handle the weapons. "Emergency preparedness at the local level appears almost nonexistent around Department of Defense and Energy [i.e., nuclear weapons] facilities," a 1979 General Accounting Office report concluded. In the event of a serious accident, the report warned, "there is only limited assurance that persons living or working near these nuclear facilities would be adequately protected."

The government will usually neither confirm nor deny that nuclear weapons are stored at a particular facility, or that nuclear weapons were involved in a particular accident. Until recently the public had no way of knowing about the risk it runs from the Pentagon's terrifying near-misses. Since the United States first began making nuclear weapons, nearly everything about them, from their design and location to the number, severity, and consequences of accidents, has been top secret—classified information. But public pressure has begun to pry information about accidents from the Pentagon. It has not, however, been prompt in coming, and it has not always been complete, or true.

One recent Broken Arrow was all too obvious to the local citizens whose lives it disrupted. This accident was widely reported in the press; perhaps as a result, it was later acknowledged officially, by the Pentagon. In September 1980, at a Titan II missile silo in a tiny community in Arkansas, a worker dropped a wrench on the missile, rupturing part of it and causing volatile fuel to spew into the silo. Several hours later, reported local residents, a blast lit up the early morning sky "like daylight." The explosion reduced the missile silo to rubble, although the nuclear warhead itself was hurled out and later recovered intact. One Air Force man was killed, 21 were badly injured, and some 1,500 civilians within a 5-mile radius of the silo were evacuated.

In spite of the fact that it ordered an evacuation, the Air Force insisted that there had been no leak of radioactive materials or other threat to residents. But in Guy, Arkansas, 5 miles from the blasted silo, a "light snow" fell about an hour after the explosion. At least two dozen residents reported nausea and a burning sensation in the nose, throat, and lungs.

Late in January 1981, four months after the Titan II accident, both Reuters news agency and reporter Stephen Talbot of KQED-TV in San Francisco obtained through the Freedom of Information Act (FOIA) a Pentagon document listing 26 other Navy and Air Force Broken Arrows. The Department of Defense had affixed the following note to the bottom of the document: "We feel confident that there are fewer than ten accidents for which we do not yet have summaries, but we cannot give an exact number at this time pending further research."

Talbot's sources could give more numbers; according to them there had been at least two other Broken Arrows in 1980 alone. One, says Talbot, was a crash of a nuclear-armed FB-111 off the coast of New England. The other was a September 15 fire aboard a B-52 carrying nuclear weapons at the Grand Forks Air Force Base in North Dakota. The information officer in charge wouldn't call this accident a Broken Arrow. "It's Air Force policy," he said, "to neither confirm nor deny the presence of nuclear weapons."

There's more evidence that the Pentagon list released in January 1981 doesn't present the whole accident story. The Navy recently acknowledged the existence of classified documents hundreds of pages long listing "Navy Nuclear Weapons Accidents and Incidents" occurring between March 1973 and March 1978.

In May 1981, in response to other FOIA requests, the Pentagon acknowledged five other Air Force accidents that occurred

The effect of the radioactivity in the food chain is compounded because of two facts. First, when radionuclides are ingested by higher life forms, they accumulate in specific organs or areas of that life form. Strontium 90, for example, is a calcium analog: that is, it tends to mimic calcium and concentrates in the bone. Cesium 137 mimics potassium and accumulates in the muscles. When the animal is devoured by a larger life form, the radionuclides concentrate in the corresponding organ of the new host.

Second, because of the nature of the food chain, the level of radioactivity can increase geometrically the higher up the chain one goes. An animal on an upper level of the food chain must consume great quantities of lower forms to survive. If one tuna eats 500 herring, 100 of which have absorbed ten ions each of cesium 137, the tuna will ingest 1,000 ions of cesium 137.

Concentration of cesium 137 has been found already in the fish caught around the Mariana Islands in the Pacific.

Mark Dowie
Mother Jones

between 1950 and 1958. In three of these the nuclear weapons involved were lost altogether.

The Army also handles thousands of nuclear weapons. So far, the public doesn't know whether the Army has had any Broken Arrows.

Nukes, or parts of them, have disappeared several times. The most mysterious incident took place over the Mediterranean in March 1956. A B-47 carrying what the Pentagon called "two capsules of nuclear weapons material" disappeared in clouds

while preparing for midair refueling. No trace of the missing crafts, crew, or nukes was ever found.

The worst near-miss in the history of Broken Arrows took place near Goldsboro, North Carolina, in January 1961.

A B-52 crashed, automatically jettisoning two 24-megaton bombs.

The Air Force maintains the bombs were "unarmed nuclear weapons." It considered them unarmed because each warhead had six interlocking safety mechanisms, all of which had to be triggered in sequence to set off the bomb. Five of the six switches in one bomb had been set off by the fall. (Couldn't the bomb just as easily have been described as "almost" or "not quite" armed?)

The second bomb broke in pieces, and although the Department of Defense says it found no trace of radioactive leakage, a part of the bomb, officially described as "a piece of the bomb containing uranium," never turned up. The Pentagon won't say how large a piece is missing, how much or what sort of uranium is involved, or even what "piece" means. "It is our opinion that since we looked so thoroughly and could not find the missing piece, the likelihood of anybody else finding it is nil," a Pentagon spokesperson offered.

January 1966 brought another particularly notorious Broken Arrow when an American B-52 carrying four hydrogen bombs collided with a refueling tanker in midair near Palomares, Spain. One bomb landed intact in a dry riverbed. One was missing for 80 days. Luckily some local fishermen had seen it fall into the sea. Eventually a manned midget submarine found this bomb on a steep slope 2,500 feet down and used mechanical arms to retrieve it.

Two of the bombs broke and released plutonium, necessitating a clean-up that cost over $50 million and the removal of 1,400 tons of contaminated soil. Not to worry. The U.S. government claimed that "with the type of radiation involved, even if a person standing in a cloud of plutonium dust inhaled the radioactive material, the resulting radiation dose would amount to less than the yearly dose permitted for absorption by U.S. atomic industry workers."

Mother Jones, the investigative reporting magazine, said the government's statement "even fifteen years ago was so misleading as to be insulting to the people of Palomares, Spain. Few scientists could argue that inhaling a 'cloud of plutonium dust' was risk free." Palomares farmers and their crops are still being monitored for radioactivity.

Two years after the Palomares incident, in January 1968, a B-52 carrying four hydrogen bombs crashed while trying to make an emergency landing at Greenland's Thule Air Force Base. Military teams on dogsleds and in helicopters spent weeks searching the ice for the missing bombs, which DOD claims were eventually found. But the council of the Eskimo village of Thule was far from happy with the bland apologies it received from the United States. According to *Mother Jones* the council "demanded to know what the U.S. meant when it assured them that there was 'probably' no radiation danger in the area. Probably the government meant it did not really know what had happened to all the radioactive material from all four bombs; probably some of it melted into the ice and snow around the Eskimo village and was swallowed by the soil of Greenland—not to return again, it is hoped."

At about the time of the Thule Broken Arrow, the Pentagon decided to make a limited admission to the continuing problem of nuclear accidents. In January 1968 it issued a press release list-

INSIDE A NUCLEAR WEAPONS PLANT, AMARILLO, TEXAS

In the administration building, visitors are greeted with a large poster showing a woman with her finger on her lips. "Shh," reads the caption, "no classified discussion here." The same poster is in the cafeteria. . . .

Although Pantex has assembled nuclear weapons since 1951, it wasn't until 1963 that its function was made public. . . .The Pantex "line" is primarily made up of bays—small, high-ceiling rooms with thick cement doors, and often heavy steel doors.

Mechanical assemblies and explosive work is done in such bays. A high explosive part in a weapon, for example, is made into a hardened form by a pressing operation.

Then it is put on a machine that resembles a potter's wheel and cut by a machine tool into the shape for the weapon involved.

In both the pressing and machining operations, the number of people on a bay is limited, normally to only one or two workers.

When the pressure is being applied to the high explosives, no one is allowed in the bay. The room is also cleared when the machining requires holes to be drilled in the explosive. The operation is run from a control room with a television camera in the bay showing how the work is going.

Security has been tightened, particularly within the weapons assemblyline area. Some of that has to do with fear of terrorists, some with last year's accident.*

Walter Pincus
The Washinton Post
December 10, 1978

**Authors' note:* An explosion at the Pantex plant on March 30, 1977, killed three workers. The accidental blast was set off while a high explosive was being machined.

ing 12 nuclear weapons accidents that were said to have taken place between February 1958 and January 1966. Until Stephen Talbot and Reuters obtained a new list, in January 1981, the 1968 press release represented the only Broken Arrows the Pentagon had ever acknowledged. Even when the 1981 list is added to the 1968 list, *the Pentagon is claiming that there were no accidents at all between the Greenland crash in 1968 and the Titan II explosion, in Arkansas, in September 1980.* Yet before 1968 the Pentagon's average was three Broken Arrows a year, with five in 1958 alone. "This dramatic improvement in the government's accident record between 1968 and 1980 may depend on how one defines *accident* and who's counting," commented *Mother Jones.* The magazine's own count—culled from news reports, journals, and unofficial sources—was a probable 138 Broken Arrows during the twelve years that the Pentagon continues to claim were accident-free.

Mother Jones is not the only observer monitoring nuclear mishaps. The Stockholm International Peace Research Institute (SIPRI), a disarmament think tank funded by the Swedish government, estimates about 125 accidents, major and minor—including two nuclear-powered submarines lost at sea—between the years 1945 and 1976. That averages out to one Broken Arrow occurring every two and a half months

Bent Spears occur even more frequently. And there are also near-catastrophes—such as a B-52 crash near a Michigan power plant in 1971, or *Mother Jones*'s count of at least 123 Titan nuclear fuel leaks in the last six years—that the Pentagon doesn't categorize as any sort of accident at all. Altogether, *Mother Jones* reports,

There have been at least 404 casualties related to nuclear weapons fires, explosions, accidents or cleanups.

Eighty-nine deaths and eighty-seven injuries have occurred due to missile or aircraft mishaps, and 228 deaths in ships lost at sea while carrying nukes.

Besides immediate injuries and deaths, nuclear accidents

have other grave consequences. The Rocky Flats weapons plant, just west of Denver, which makes plutonium parts for nuclear weapons, has had over 200 fires and other officially recorded accidents. In May 1969 one of the worse fires in the history of the nuclear industry consumed hundreds of pounds of plutonium. The cleanup alone cost taxpayers $45 million.

In 1974 Dr. Carl Johnson, then the health director of Jefferson County, Colorado, where Rocky Flats is located, began testing soil near the weapons facility and found that the soil contained up to 3,390 times the level of radioactive plutonium than was expected. In 1979 Johnson released another study that found abnormally high cancer rates among people living downwind from Rocky Flats.

Dr. Johnson ran into opposition from local land developers who wanted to build on contaminated land, and was recently fired from his job, but not before a National Cancer Institute study bore out his findings. Conducted between 1969 and 1971, the study showed that in Colorado communities downwind of Rocky Flats, cancer rates were 24 percent higher than normal for men and 10 percent higher for women. Johnson has noted, "The higher incidence of cancer in the areas near the plant were due to more cases than expected of leukemia, lymphoma, and myeloma and cancer of the lung, thyroid, breast, esophagus, stomach and colon, a pattern similar to that observed in the survivors of Hiroshima and Nagasaki." Johnson believes workers at facilities like Rocky Flats and citizens living near them both run very high risks of chromosomal damage and cancer. A National Academy of Sciences committee recently agreed when it published the following estimate:

One person in a thousand will develop cancer from radiation by the year 2000.

The rates among nuclear workers will be higher.

There are additional weapons-related health hazards that the Pentagon has yet to acknowledge. Over a thousand residents of the Nevada test site where 93 atom bombs were tested in the desert between 1952 and 1963 (at which time the nuclear test-ban treaty became effective) have sued the federal government for $500 million. The Atomic Energy Commission, forerunner of what is now the Nuclear Regulatory Commission, carried out testing at the site only when the winds were blowing away from Las Vegas, so that the fallout would pass, instead, over sparsely populated desert areas. The residents in those sparsely populated areas noted a definite increase in cases of leukemia, birth defects, sterility, miscarriage, mental retardation, and cancer. Also, thousands of sheep and other range animals died mysteriously.

Throughout the testing period the AEC insisted that none of these misfortunes was caused by fallout. A 1955 AEC pamphlet praised the patriotic, conservative, largely Mormon "downwind people." "Some of you have been inconvenienced by our test operations," the AEC observed, patronizingly.

"At times some of you have been exposed to potential risks from flash, blast, or fallout. You have accepted the inconvenience or the risk without fuss, without alarm, and without panic. Your cooperation has helped achieve an unusual record of safety."

An article in *Life* in 1980 noted that medical reports have confirmed what the downwind residents suspected all along. *Children in southern Utah born during the heaviest testing were 2.4 times more likely to die of leukemia than children born before or since.*

Hearings in the downwind residents' suit began in the fall of 1981 in Salt Lake City. Because there are so many plaintiffs the case is expected to go on for years. Among their lawyers is Stewart Udall, former Secretary of the Interior. The downwind people are attempting to prove that the Atomic Energy Commission knew enough about the consequences of nuclear weapons testing to foresee the risks, and that the residents had not been adequately warned. (So far, no one has ever successfully sued the government on a health-related radiation claim.)

"I lived here since I was about two years of age. I watched the Strategic Air Command develop [indistinct] what does this mean? Ten thirty at night a Looking Glass takes off. You think about it. You know what Looking Glass is, you know they're going up and that's their mission. To retaliate in case of a strike against this country so yes, it's like on a daily basis, you can't dwell on it, you have to dwell on other things, but it, it's there, and you always know it's there."

Jerry Allen
Resident (along with his wife and two children) of Omaha, Nebraska, home of SAC headquarters

CBS Reports
Sunday, June 14, 1981

NUCLEAR WASTE: ACCIDENT OR INEVITABILITY

Between 1946 and 1970 barges and planes dropped radioactive trash from military and domestic nuclear plants into 50 ocean dumps up and down the East and West coasts. The largest dumps are but a few hours by boat from New York, Newark, Boston, Los Angeles, and San Francisco, in prime fishing grounds. Nearly 50,000 concrete barrels full of nuclear waste lie about 23 miles out of San Francisco's harbor. Many barrels have cracked under the ocean's tremendous pressure. Marine scientists believe that fish are eating the radioactive waste. Concentrations of cesium-137 have been found already in fish caught near the Mariana Islands in the Pacific. Giant mutant sponges, an early warning of future hazards for humans, have turned up not far from San Francisco.

"You couldn't actually *design* a better way to put that radioactivity into our food," says Jackson Davis, of the psychobiology department of the University of California.

Davis expects peak release of radioactivity from the dumps during the 1980s. (Actually, in comparison to other countries, the United States has dumped a relatively modest amount of radioactive waste: 11,000 tons. [This figure includes an entire 30-feet-high nuclear reactor vessel jettisoned from a nuclear sub 120 miles offshore at the Delaware-Maryland border in 1959 and never recovered.] Western European nations dumped some 65,000 tons of radioactive waste into the Irish Sea between 1967 and 1979. Japan, which is running out of land dumps, plans to drop 10,000 barrels into the Pacific annually, beginning in 1982.)

The Pentagon cites "national security" as its reason for failing to inform the public about waste dumping, weapons testing, or the location of nuclear weapons. Retired Rear Admiral Gene La Rocque, a former assistant director of stategic planning in the office of the Chief of Naval Operations, now heads the Center for Defense Information, an independent research institute in Washington, D.C. "Everyone ought to know where the nuclear weapons are," he says.

"The reason the military does not tell people whether or not there are weapons on a ship or stored in a certain place is they are afraid that people will be unhappy about it and want those weapons removed."

Peace and antinuclear groups are attempting to identify 120 to 200 weapons installations around the country and to broadcast the issue of public safety. Facilities storing "special weapons" (Pentagon jargon for nuclear weapons) are relatively easy to spot. They have double fencing, 24-hour lighting, posted signs authorizing "use of deadly force" against trespassers, Marine or other special guards, and "clear" zones. Some activists in Hawaii have sued to compel the Navy to file an environmental impact statement on the possible hazards of one nuclear-weapons storage site there. A lawyer with the Center for Constitutional Rights in New York believes that the Hawaiian suit may eventually force the Pentagon to disclose the risks of all other sites.

We Americans know little of nuclear accidents caused by U.S. weapons, and even less about mishaps in other countries. Given the incomplete information released by the government in

U.S. NUCLEAR WEAPONS ACCIDENTS (1950–80) ACKNOWLEDGED BY THE PENTAGON

Date	Weapon System	Location	Description*
1. February 13, 1950	B-36	Pacific Ocean, off Puget Sound, Washington	B-36 on simulated combat mission dropped nuclear weapon from 8,000 feet over Pacific Ocean before crashing. "Only the weapon's high explosive material detonated. . . . A bright flash occurred on impact, followed by a sound and shock wave."
2. April 11, 1950	B-29	Manzano Base, New Mexico	B-29 crashed into a mountain, killing the crew, three minutes after takeoff from Kirtland AFB. Bomb case demolished, high explosive burned, nuclear components recovered.
3. July 13, 1950	B-50	Lebanon, Ohio	B-50 on training mission from Biggs AFB, Texas, crashed, killing sixteen crewmen. "The high explosive portion of the nuclear weapon aboard detonated on impact."

4. August 5, 1950	B-29	Fairfield-Suisun AFB, California (now Travis AFB)	B-29 carrying a nuclear weapon crashed on takeoff and the high explosive material detonated. Nineteen crewmen and rescuers killed, including General Travis.
5. March 10, 1956	B-47	Mediterranean Sea	B-47 carrying "two capsules of nuclear weapons material" from MacDill AFB to an overseas base disappeared in clouds while preparing for midair refueling. "An extensive search failed to locate any traces of the missing aircraft or crew."
6. July 26, 1956	B-47	Overseas Base	B-47 with no weapons on board crashed into "a storage igloo containing several nuclear weapons. The bombs did not burn or detonate."
7. May 22, 1957	B-36	Kirtland AFB, New Mexico	B-36 ferrying a nuclear weapon from Biggs AFB, Texas, to Kirtland dropped the bomb in the New Mexico desert. "The high explosive material detonated, completely destroying the weapon and making

U.S. NUCLEAR WEAPONS ACCIDENTS (1950–80) ACKNOWLEDGED BY THE PENTAGON (*Continued*)

Date	Weapon System	Location	Description*
			a crater approx. 25 feet in diameter and 12 feet deep . . . no radioactivity beyond the lip of the crater."
8. July 28, 1957	C-124	Atlantic Ocean	C-124 aircraft en route from Dover AFB, Delaware, lost power in two engines and jettisoned two nuclear weapons over the ocean. "No detonation occurred from either weapon." The nuclear weapons were never found. The plane landed near Atlantic City, N.J.
9. October 11, 1957	B-47	Homestead AFB, Florida	B-47 crashed shortly after takeoff with a nuclear weapon and a nuclear capsule on board. "Two low order detonations occurred during the burning."
10. January 31, 1958	B-47	Overseas Base	B-47 crashed and burned during takeoff with one nuclear weapon

			"in strike configuration. . . . There was some contamination in the immediate area of the crash."
11. February 5, 1958	B-47	Savannah, Georgia (Hunter AFB)	B-47 on simulated combat mission out of Homestead AFB, Florida, had a midair collision with an F-86 aircraft. The B-47 jettisoned the nuclear weapon "five miles southeast of the mouth of the Savannah River (Georgia) in Wassaw Sound off Tybee Beach." The weapon was never found.
12. March 11, 1958	B-47	Florence, South Carolina	B-47 left Hunter AFB, Georgia, with three other B-47s en route to an overseas base. "The aircraft accidentally jettisoned an unarmed nuclear weapon, which impacted in a sparsely populated area $6\frac{1}{2}$ miles east of Florence, South Carolina. The bomb's high explosive material exploded on impact. The detonation caused property damage but no injuries on the ground."

U.S. NUCLEAR WEAPONS ACCIDENTS (1950–80) ACKNOWLEDGED BY THE PENTAGON (*Continued*)

Date	Weapon System	Location	Description*
13. November 4, 1958	B-47	Dyess AFB, Texas	B-47 caught fire on takeoff and crashed, killing one crew member. The high explosive in the nuclear weapon on board exploded, leaving "a crater 35 feet in diameter and six feet deep."
14. November 26, 1958	B-47	Chennault AFB, Louisiana	B-47 caught fire on the ground. Single nuclear weapon on board was destroyed by fire. "Contamination was limited to the immediate vicinity."
15. January 18, 1959	F-100	Pacific Base	F-100 caught fire on the ground while loaded with unarmed nuclear weapon.
16. July 6, 1959	C-124	Barksdale AFB, Louisiana	"A C-124 on a nuclear logistics movement mission crashed on takeoff. The aircraft was destroyed by fire, which also destroyed one

			weapon. . . . Limited contamination . . ."
17. October 15, 1959	B-52 & KC-135	Hardinsberg, Kentucky (Glen Bean, Kentucky)	B-52 and KC-135 operating out of Columbus AFB, Mississippi, collided during refueling over Kentucky. DOD summary provides no information about nuclear weapons, but previous reports said two unarmed nuclear weapons were recovered undamaged.
18. June 7, 1960	Bomarc missile	McGuire AFB, New Jersey	"A Bomarc air defense missile in ready storage condition [permitting launch in two minutes] was destroyed by explosion and fire after a high pressure helium tank exploded and ruptured the missile's fuel tanks. The warhead was also destroyed by the fire although the high explosive did not detonate. Contamination was restricted to an area . . . approx. 100 ft. long."
19. January 24, 1961	B-52	Goldsboro, NC	B-52 crashed during airborne alert

U.S. NUCLEAR WEAPONS ACCIDENTS (1950–80) ACKNOWLEDGED BY THE PENTAGON (*Continued*)

Date	Weapon System	Location	Description*
			mission and dropped two nuclear weapons near Goldsboro, N.C. "A portion of one weapon, containing uranium, could not be recovered despite excavation in the waterlogged farmland to a depth of 50 feet. The Air Force subsequently purchased an easement requiring permission for anyone to dig there. There is no detectable radiation and no hazard in the area." The DOD summary does not mention the fact that five of six interlocking safety triggers on the bomb failed. "Only a single switch," reported Dr. Ralph Lapp, head of the nuclear physics branch of the Office of Naval Research, "prevented the 24-megaton bomb from detonating and spreading fire and destruction over a wide area."

20. March 14, 1961	B-52	Yuba City, California (Beale AFB)	B-52 with two nuclear weapons on board crashed after pilot succeeded in "steering the plane away from a populated area." No high explosive detonation and no contamination. Safety devices worked.
21. January 13, 1964	B-52D	Cumberland, Maryland	B-52D en route from Westover AFB, Massachusetts, to its home base at Turner AFB, Georgia, crashed with two unarmed nuclear weapons on board, which were recovered "relatively intact."
22. December 5, 1964	LGM 30B (Minuteman ICBM)	Ellsworth AFB, South Dakota	LGM 30B Minuteman I missile was on strategic alert at Launch Facility L-02, Ellsworth AFB, South Dakota, when a "retrorocket" accidentally fired during repairs. There was considerable damage but "no detonation or radioactive contamination."
23. December 8, 1964	B-58	Bunker Hill AFB (now Grissom AFB), Indiana	B-58 crashed while preparing for takeoff on icy runway at Bunker Hill AFB, Indiana. "Portions of the nuclear weapon burned;

U.S. NUCLEAR WEAPONS ACCIDENTS (1950–80) ACKNOWLEDGED BY THE PENTAGON (*Continued*)

Date	Weapon System	Location	Description*
			contamination was limited to the immediate area of the crash and was subsequently removed."
24. October 12, 1965	C-124	Wright-Patterson AFB, Ohio	Fire during refueling on ground burned "components of nuclear weapons and a dummy training unit. . . . The resultant radiation hazard was minimal."
25. January 17, 1966	B-52 & KC-135	Palomares, Spain	B-52 and KC-135 collided during midair refueling and both aircraft crashed near Palomares, Spain. The B-52 carried four nuclear weapons. One was recovered on the ground and one was finally found in the sea after an intensive, four-month search. "Two of the weapons' high explosive materials exploded on impact with the ground, releasing some radioactive materials. Approximately 1,400 tons of

				slightly contaminated soil and vegetation were removed to the United States for storage at an approved site." The DOE has reported that the cleanup operation has cost $50 million and that the Palomares area is still being monitored for radiation today.
26.	January 21, 1968	B-52	Thule, Greenland	B-52 from Plattsburgh AFB, New York, crashed and burned some seven miles southwest of the runway at Thule AFB, Greenland. "The bomber carried four nuclear weapons, all of which were destroyed by fire. . . . Some 237,000 cubic feet of contaminated ice, snow and water, with crash debris, were removed to an approved storage site in the United States over the course of a four-month operation."
27.	September 19, 1980	Titan II ICBM	Damascus, Arkansas	During routine maintenance in a Titan II missile silo, an Air Force

U.S. NUCLEAR WEAPONS ACCIDENTS (1950–80) ACKNOWLEDGED BY THE PENTAGON (*Continued*)

Date	Weapon System	Location	Description*
			repairman dropped a socket wrench, which punctured a fuel tank and caused an explosion. One Air Force man was killed, 21 injured. The nuclear warhead, which was hurled from the silo by the explosion, was recovered intact. Local residents were evacuated. "There was *no radioactive* contamination."

Released by Stephen Talbot

*An accompanying DOD memo states: "Due to differences in record-keeping among the Services, it is difficult to say that this list is complete, particularly with respect to the earlier period covered by the summaries. The Navy and Air Force may have experienced a few more accidents and we are researching this possibility. The Army reports that it has never had an accident with nuclear weapons comparable in seriousness with those on this list. We feel confident that there are fewer than 10 accidents for which we do not yet have summaries, but we cannot give an exact number at this time pending further research. . . . This list does not contain *every* event reportable under safety regulations as an 'accident/incident involving nuclear weapons.'. . . Criteria for inclusion in this summary included the severity of the event, e.g., an airplane crash, radioactive contamination, weapon actually involved in the event, etc."

the past, it's perfectly possible that a Broken Arrow has already caused a nuclear detonation. It's clear that nuclear accidents occur frequently, and as more complicated weapons systems are developed, the chances of something going wrong multiply. Nuclear critics believe that a major accident is inevitable without a sharp reduction in the number of nukes in the U.S. arsenal, a drastic improvement in safety and handling procedures, and much greater public awareness of the problem. Says Admiral La Rocque: *"When we first had nuclear weapons, and when I had them in San Francisco harbor driving around with my cruiser, people were very intense and nervous about having nuclear weapons on board, so safety precautions were rigidly observed and our men were well trained. But as time has gone on, and I've watched it happen, we've become more lax. We have come to accept nuclear weapons as conventional weapons. We have become careless. We have trained a lot of people, but they don't stay with us very long. So we have high school, less than high school, graduates taking care of our nuclear weapons. I think there is a great danger in a harbor like San Francisco that one of these days there is going to be a collision of ships in the fog, and we are going to scatter nuclear weapons all over the harbor."*

La Rocque's nightmare is not even a "worst case scenario."

The ultimate nuclear accident would be nuc-flash—war as mistake.

One country could detonate a bomb on a neighbor's territory by accident, or accidentally drop a bomb on its own territory and then blame a neighbor. The possibility of accidental war ("Who's minding the button?") crossed people's minds in the wake of President Reagan's near-assassination in April 1981.

Five months earlier, in December 1980, during an exercise in mobilization conducted by the Pentagon and 35 other federal agencies, there was a twelve-hour breakdown of the worldwide command-and-control computer system at the very peak of the simulated crisis. (See chapter 10.) Programmers spent six hours finding the right code sequence to secure up-to-date readiness information.

It is painfully clear that U.S. taxpayers have little security for the $2.3 trillion they have spent on the Pentagon since the end of World War II. We not only pay for weapons; we also bear the cost of cleanup after accidents. Movement of weapons and weapons material increases the incidence of accidents—train and truck accidents have spilled uranium in North Carolina, Colorado, and Kansas. As Stephen Talbot has pointed out:

"U.S. arms have never killed a Soviet soldier, but they've killed and injured many Americans, polluted our environment, and possibly caused an increase in cancer."

Since the beginning of the nuclear age, the civilian and military programs combined have created over 76 million cubic feet of high-and-low level wastes, and over 140 million tons of radioactive uranium mine and mill tailings. These wastes are now temporarily housed at hundreds of sites around the country. At the Hanford Nuclear Reservation in eastern Washington, over 430,000 gallons of high-level military wastes have leaked from sixteen different storage tanks.

Pam Solo and Mike Jendrzejczyk
Business and Society Review
Spring 1980, No. 33

NUCLEAR WEAPONS STORAGE AND DEPLOYMENT SITES

ARMY

- ✲ arsenal, very likely storage
- ✱ training with nuclear capable weapons

NAVY

- ○ nuclear capable surface ships
- ◇ nuclear attack submarine
- △ ballistic missile submarine
- ▫ nuclear capable planes
- ✧ testing/storage site

AIR FORCE

- ▲ Minuteman II, III
- ◆ Titan II, III
- ✦ B-52
- ● F-111 and FB-111
- ◒ F-101 and F-106
- ★ other nuclear capable planes
- ▪ storage facility likely

DEPARTMENT OF ENERGY

- ❋ production, testing, or storage

(*Note:* This information and map come from NARMIC [National Action/Research on the Military Industrial Complex.] A Project of the American Friends Service Committee. Sources used were House and Senate hearings, *Army* magazine, *Air Force Magazine*, and publications of the Stockholm International Peace Research Institute and the Center for Defense Information.)

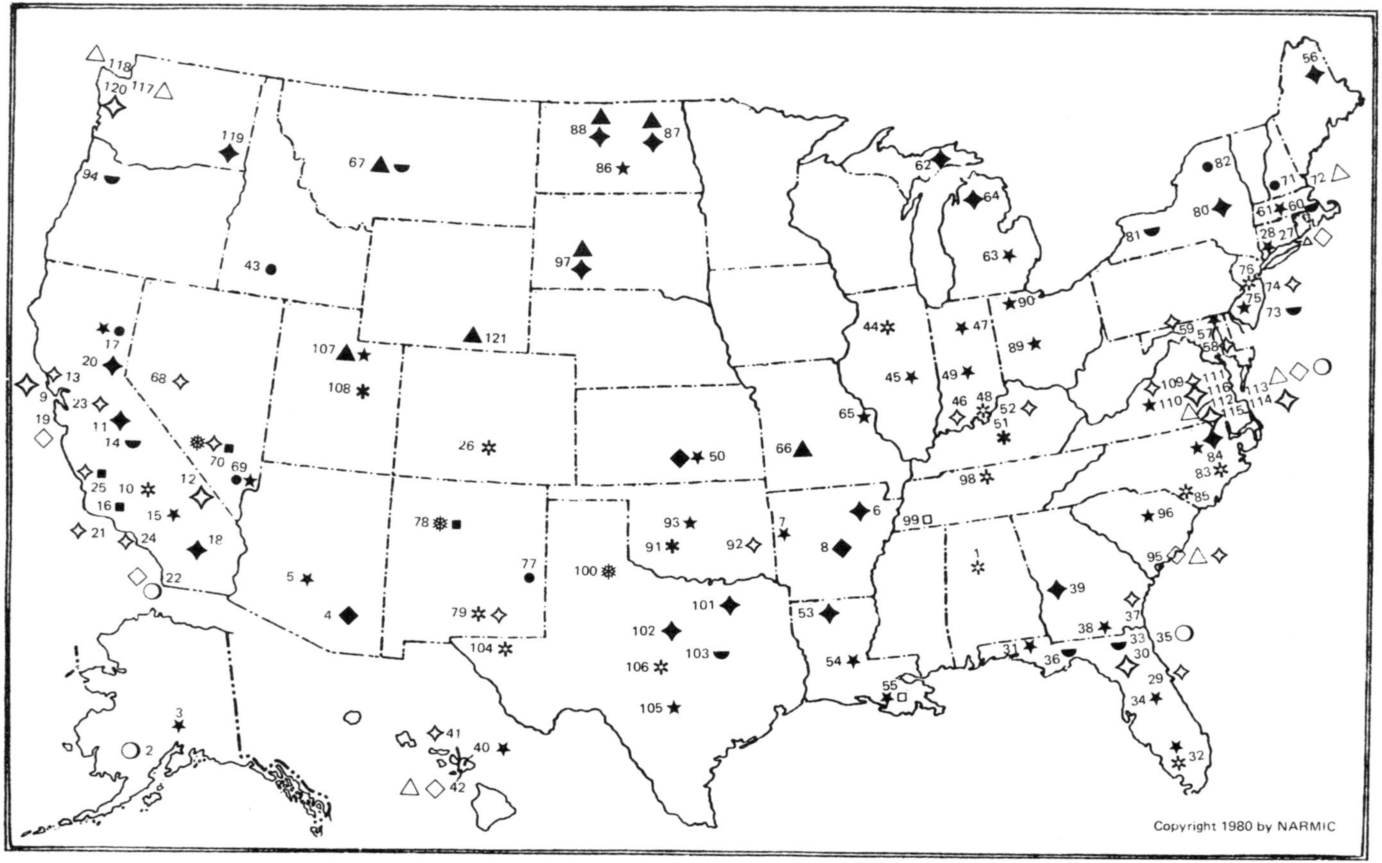

NUCLEAR WEAPONS AND DEPLOYMENT SITES

State	Map Number	Installation-Location	Delivery Systems or Type of Facility
Alabama	1	Redstone Arsenal	Missile and munitions center and school
Alaska	2	Adak Naval Station	Nuclear-capable Naval surface ships
	3	Elmendorf AFB	F-4
Arizona	4	Davis-Monthan AFB	Titan II
	5	Luke AFB	F-4
Arkansas	6	Blytheville AFB	B-52
	7	Fort Smith	F-100D
	8	Little Rock AFB	Titan II
California	9	Alameda NAS	Naval weapons
	10	Angeles National Forest	Army 19th Artillery weapons storage
	11	Castle AFB	B-52
	12	China Lake NWS	Naval weapons station
	13	Concord NWS	Naval weapons station
	14	Fresno	F-106

NUCLEAR WEAPONS AND DEPLOYMENT SITES (*Continued*)

State	Map Number	Installation-Location	Delivery Systems or Type of Facility
	15	George AFB	F-4 C/E/G
	16	Los Angeles AFB	Development, production, delivery of ballistic missiles
	17	McClellan AFB	F-111, FB-111, F-100, F-105
	18	March AFB	B-52
	19	Mare Island Naval Shipyard (Vallejo)	5 attack subs
	20	Mather AFB	B-52
	21	Point Mugu NAS	Pacific missile test center
	22	San Diego Navy Base	3 guided missile cruisers, 12 attack subs
	23	San Francisco Bay Naval Shipyard	Missile testing
	24	Seal Beach NWS	ASROC, SUBROC, TALOS (Naval weapons)
	25	Vandenberg AFB	ICBM test facilities (Navy and Air Force)
Colorado	26	Fort Carson	Army nuclear weapons storage

Connecticut	27	Groton/New London Naval Submarine Base (Thames River)	20 attack subs, ballistic missile subs
	28	Windsor Locks	F-100
Florida	29	Cape Canaveral	Trident testing
	30	Cecil Field NAS	Naval weapons
	31	Eglin AFB, Hulbert Field	F-4E
	32	Homestead AFB	F-4E, Nike Hercules
	33	Jacksonville NAS	F-106
	34	MacDill AFB	F-4D/E, F-16
	35	Mayport NS	Nuclear-capable surface ships
	36	Tyndal AFB	F-106; air defense weapons center
Georgia	37	King's Bay (Navy)	Sub support base
	38	Moody AFB	F-4E, F-4
	39	Robins AFB	B-52
Hawaii	40	Hickam AFB	F-4
	41	Laulaule Naval Magazine	Warheads for ballistic missile subs and attack subs
	42	Pearl Harbor Naval Submarine Base	14 attack subs, 10 ballistic missile subs

NUCLEAR WEAPONS AND DEPLOYMENT SITES (*Continued*)

State	Map Number	Installation-Location	Delivery Systems or Type of Facility
Idaho	43	Mountain Home AFB	F-111A
Illinois	44	Rock Island Arsenal	Army munitions
	45	Springfield	F-4C
Indiana	46	Crane NWSC	Naval weapons
	47	Fort Wayne	F-4C
	48	Indiana AAP	155-mm howitzers, 8-inch howitzers
	49	Terre Haute	F-100D
Kansas	50	McConnell AFB	F-105, F-4, Titan II
Kentucky	51	Louisville (Army National Guard) Corps Artillery	Honest John (MGRIB)
	52	Louisville Ordnance Station (Navy)	Naval weapons
Louisiana	53	Barksdale AFB	B-52
	54	England AFB	A-7
	55	New Orleans NAS	F-100, F-4C

Maine	56	Loring AFB	B-52
Maryland	57	Andrews AFB	F-105D, F-4
	58	Indian Head Naval Ordnance Station	Naval weapons
	59	White Oak Naval Surface Weapons Center	Naval weapons
Massachusetts	60	Otis AFB	F-106
	61	Westfield	F-100D
Michigan	62	K. I. Sawyer AFB	B-52
	63	Selfridge ANGB	F-4C/D
	64	Wurtsmith AFB	B-52
Missouri	65	St. Louis	F-4C
	66	Whiteman AFB	Minuteman II
Montana	67	Malmstrom AFB (Great Falls)	F-106, Minuteman II, III
Nevada	68	Hawthorne Naval Ammunition Depot	Naval weapons
	69	Nellis AFB	F-4D/E, F-111E/F, F-16
	70	Nevada Test Site	Testing by DOE and all military branches

NUCLEAR WEAPONS AND DEPLOYMENT SITES (*Continued*)

State	Map Number	Installation-Location	Delivery Systems or Type of Facility
New Hampshire	71	Pease AFB	FB-111
	72	Portsmouth	Ballistic missile subs
New Jersey	73	Atlantic City	F-106
	74	Earle Naval Weapons Station	Naval weapons
	75	McGuire AFB	F-105B
	76	Picatinny Arsenal	Radiological testing of weapons
New Mexico	77	Cannon AFB	F-111D
	78	Kirtland AFB (Manzano)	Nuclear weapons laboratory storage of "obsolete" weapons
	79	White Sands Missile Range	Missile testing
New York	80	Griffiss AFB	B-52
	81	Niagara Falls International Airport	F-101
	82	Plattsburgh AFB	FB-111
North Carolina	83	Fort Bragg (Fayetteville)	Army munitions
	84	Seymour-Johnson AFB	F-4F, B-52

	85	Sunny Point Military Ocean Terminal	Ammunition shipping point
North Dakota	86	Fargo	F-4C/D
	87	Grand Forks AFB	B-52, Minuteman III
	88	Minot AFB	B-52, Minuteman III
Ohio	89	Springfield	F-100, A-7
	90	Toledo	F-100D
Oklahoma	91	Fort Sill	Lance missiles
	92	McAlester Naval Ammunition Depot	Naval weapons
	93	Tinker AFB	F-105D/F
Oregon	94	Portland (ANG)	F-101
South Carolina	95	Charleston Fleet Ballistic Missile Submarine Center	5 attack subs, Polaris subs, ballistic missile subs (also will have Trident)
	96	Shaw AFB	F-4, F-4D
South Dakota	97	Ellsworth AFB	B-52, Minuteman II
Tennessee	98	Clarksville (Army)	Army munitions
	99	Memphis NAS	9 P-3A's

NUCLEAR WEAPONS AND DEPLOYMENT SITES (*Continued*)

State	Map Number	Installation-Location	Delivery Systems or Type of Facility
Texas	100	Amarillo (Pantex)	(DOE) final assembly of nuclear weapons
	101	Carswell AFB	B-52
	102	Dyess AFB	B-52
	103	Ellington AFB (ANG)	F-101
	104	Fort Bliss	Nike Hercules
	105	Kelly AFB	F-4C
	106	Killeen	Defense Atomic Support (weapons storage)
Utah	107	Hill AFB	Minuteman silo, F-4D, F-16, F-105D
	108	Salt Lake City (Army National Guard) Field Artillery Unit	Honest John unguided artillery rocket
Virginia	109	Advanced Undersea Weapons Compound (exact location not known)	Naval nuclear weapons storage
	110	Byrd Field (Sandston)	F-105D

	111	Dahlgren Navy Surface Weapons Lab	Naval weapons storage
	112	Newport News	Ballistic missile sub
	113	Norfolk Naval Station	Ballistic missile, attack, Polaris subs, and nuclear-capable surface ships
	114	(Oceana NAS) Virginia Beach	Naval weapons storage, testing of MK 88, MOD-1
	115	St. Juliens Cr. NWS	Naval weapons storage
	116	Yorktown NWS	Naval weapons storage
Washington	117	Bangor	Trident
	118	Bremerton Polaris Missile Facility	Polaris, Poseidon
	119	Fairchild AFB	B-52
	120	Keyport Torpedo Station	Torpedos
Wyoming	121	F. E. Warren AFB	Minuteman III

Credit: Rebekah Ray-Crichton for research and preparation of map and accompanying material

August 6, 1945: atomic burst. At the time this photo was made, smoke billowed 20,000 feet above Hiroshima while smoke from the burst of the first atomic bomb had spread over 10,000 feet on the target, at the base of the rising column. Two planes participated in this mission, one to carry the bomb, the other to act as escort.

chapter 16

After the Blast

Physicians and scientists in the U.S., the Soviet Union, Japan, and European countries have joined together in recent years in a systematic effort to debunk the dangerous myth that nuclear war can be fought, survived, and even won. Dr. Howard H. Hiatt of Harvard's School of Public Health refers to nuclear war as "the last epidemic." He and other prominent scientists argue that neither the public nor politicians have a realistic grasp of the biological effects of nuclear war on its survivors, of the economic, social, and ecological devastation it would incur. In short, we do not recognize that in the wake of a nuclear war these 38 years after Hiroshima, civilization as we know it would not survive.

Doctors and scientists maintain that civil defense plants are useless because the blast's pressure, heat, and radiation would kill even people in shelters, and fallout would soon reach evacuees. They consider medical disaster planning for nuclear war meaningless because hospitals would be destroyed, and medical personnel injured or killed. Any hospitals, doctors, and supplies that remained would be insufficient for coping with the unprecedented numbers of burns, broken bones, and cases of shock and radiation sickness.

Believing that prevention is the only cure for nuclear war, some Harvard biologists formed a study group to lay out the biological effects of a nuclear explosion.* In terms of the millions of deaths a nuclear war would cause, such a disaster is unthinkable—virtually beyond comprehension. But the specific ways a nuclear war would kill or maim living things can be laid out quantitatively in scientific terms.

*The original Biology of Nuclear War Study Group members were Nora Chapman, Ann Howard, Gary Howard, Neocles Leontis, Jonathan Logan, Steven Orzack, Helen Sang, and Scott Thacher.

■ The bombs dropped on Hiroshima and Nagasaki were like child's toys compared to the powerful hydrogen weapons now owned by the Soviets and the United States.

Any one of these could produce at least a thousand times the lethal blast, heat, and radiation unleashed on the Japanese cities.

It would take 909 Fat Boys, the 22-kiloton plutonium bomb that destroyed Nagasaki, to equal just one of today's 20-megaton bombs. Some strategic warheads equal 8 billion tons of TNT, or 600,000 Little Boys, the 13-kiloton bomb that fell on Hiroshima. To keep these figures in perspective, bear in mind that all of the old-style bombs the Allies dropped during World War II equaled only about 2 megatons of TNT.

As the mushroom clouds from a modern nuclear weapon rose into the stratosphere, it would carry millions of radioactive particles that would spread over the northern hemisphere within months. During the next few years the particles would fall to earth, carrying radioactive elements such as carbon-14 and strontium-90, which can be incorporated into the human body. Eventually even those who were nowhere near the blast would be exposed to 5 to 10 rads of radiation, in addition to the background radiation we are all exposed to every year from natural sources such as the sun.

■ The unfortunate city selected as target for the 20-megaton bomb would be flattened as far as 8 miles from the blast center. Nothing living would remain. Those near the center of the explosion would be vaporized within a fraction of a second, incinerated by the white flash, or blown apart or into or under something by the explosion. Those farther away from the hypocenter would be crushed by collapsing

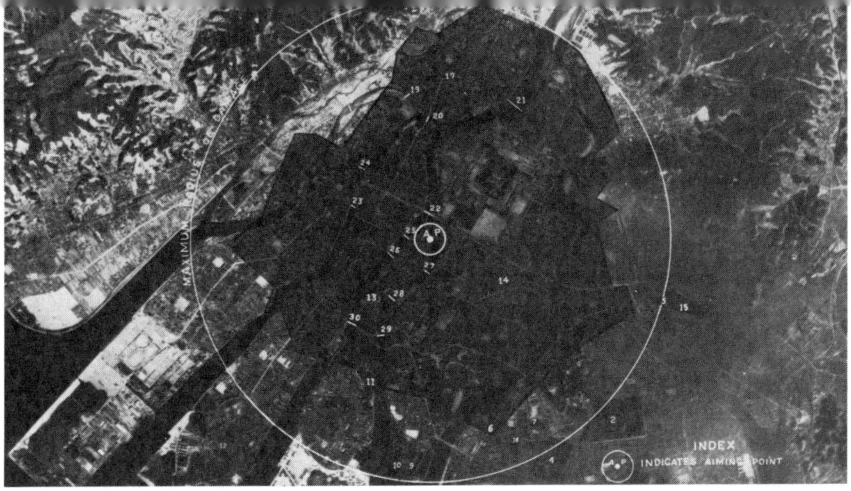

The total area devastated by the atomic strike on Hiroshima is shown in the darkened area (within circle) of photograph. Numbered items are military and industrial installations with percentages of total destruction.

1.	*Army Transport Base*	*25%*
2.	*Army Ordnance Depot*	
3.	*Army Food Depot*	*35%*
4.	*Army Clothing Depot*	*85%*
5.	*E. Hiroshima RR station*	*30%*
6.	*U/E Industry*	*90%*
7.	*Sumitomo Rayon Plant*	*25%*
8.	*Kinkwa Rayon Mill*	*10%*
9.	*Teikoku Textile Mill*	*100%*
10.	*Power Plant*	*?*
11.	*Oil Storage*	*on fire*
12.	*Electric Power Station*	*100%*
13.	*Electric Power Generator*	*100%*
14.	*Telephone Company*	*100%*
15.	*Hiroshima Gas Works*	*100%*
16.	*Hiroshima RR Station*	*100%*
17.	*Railroad Station U/E*	*100%*
18.	*Bridge*	*intact*
19.	*Bridge*	*¼ missing*
20.	*Large bridge, shattered*	*¼ missing*
21.	*Bridge, large hole in side*	
22.	*Bridge, banks caved in*	
23.	*Bridge, debris covered*	
24.	*Both bridges intact*	
25.	*Bridge*	*100%*
26.	*Bridge, severely damaged*	
27.	*Bridge, destroyed*	
28.	*Bridge, shattered, inoperative*	
29.	*Bridge, slight damage*	
30.	*Bridge, severely damaged*	

Source: *Air Intelligence Reports*

buildings, lacerated by flying splinters of wood or glass, or charred by the firestorms that would follow the explosion.

At Hiroshima and Nagasaki, people 2 miles away survived; with one of today's bombs they would be vaporized or charred within a fraction of a second after the fireball; 3 miles away sheet metal and glass would be vaporized. People 8 miles from the explosion would be severely burned; their eyeballs would melt from the heat. The heat would be followed by a soundless blast wave that would smash everything standing. Then the firestorms would begin, huge whirlwinds of fire that would rage out of control for hours, creating strong winds and sucking up all the oxygen in the air so that people would drop in their tracks, asphyxiated. Gas tanks would explode and fuel the fires. Radiation particles would blanket the landscape or be sucked up into the colder air overhead, where they would condense and drop to earth again in the form of an oily black rain filled with fallout.

■ The detonation that produces such intense pressure, powerful heat, and widespread radiation begins with a chemical explosion of a conventional explosive like TNT. The explosion causes the plutonium nuclei in the bomb to fission, causing fusion of hydrogen and lithium nuclei, which releases tremendous amounts of energy. This great pressure and heat, hot as the sun's center, triggers more fissioning of plutonium or uranium. The fission produces radiation in several forms—alpha particles (helium nuclei), beta particles (electrons), gamma rays and X rays, neutrons, and radionuclides. As the surrounding air absorbs them some of these particles and rays produce heat as high as several tens of millions of degrees Celsius. The heated air expands to produce the blast, and the great heat is radiated long-distance as infrared and ultraviolet light capable of burning exposed skin miles from the actual detonation. Radioactive particles spread throughout the blast area and the upper atmosphere.

■ The effects of fallout and radioactivity take many unpleasant and insidious forms. Bernard T. Feld, editor-in-chief of the *Bulletin of Atomic Scientists,* estimates that a major Soviet attack against American missiles and military facili-

This child from Bikini, in the Marshall Islands, has radiation burns. Bikini was used for atmospheric nuclear bomb tests from 1946 until 1958. On March 1, 1954, the United States conducted its largest bomb test ever. "Soon after the tests," according to Marshall Islander John Anjian, "people had nausea, and they had burns on their skins, and their hair fell out. On the third day the U.S. ship came and evacuated the island."

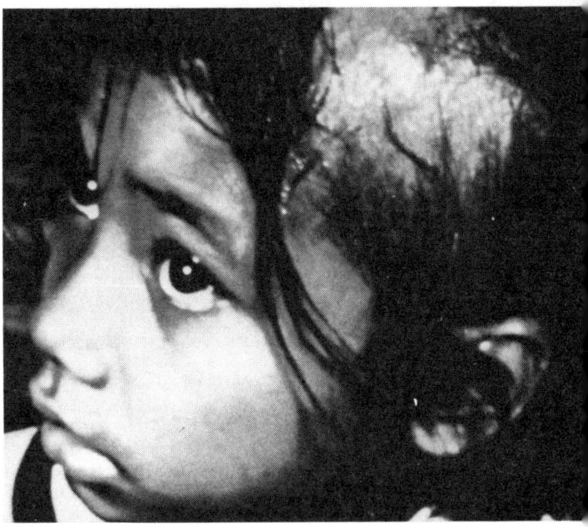

ties—a "limited war"—as defense officials currently like to imagine it—would contaminate 5 million square miles with radioactivity. In short, the entire United States. Even if 100 million people survived the blast in shelters, they wouldn't live long. They would inevitably inhale radionuclides, which would also land on the skin and burn it.

RAD POWER

A large dose of radioactivity, say about 10,000 rads, will cause death within hours or days due to the breakdown of blood vessels. After a dose of 1,000 rads the cells lining the gut are destroyed; the victim dies within a week or two. According to the Biology of Nuclear War Study Group, the thermal energy in 500 rads is only enough to warm a cup of water a few degrees Celsius, yet it can kill one out of two people after two weeks or more, because it destroys bone marrow and white blood cells. If the dead white cells go unreplaced, the victims die.

In Hiroshima and Nagasaki 88 percent of the population within a 2-mile radius of the epicenter died either instantly or within a few weeks from burns, internal injuries, and radiation sickness. Survivors' drawings of the aftermath graphically depict the long black strips of skin they saw hanging from the arms and torsos of the living, and the charred bodies and melted hands and faces of the dead. By November 1945, 130,000 were dead in Hiroshima and roughly 70,000 in Nagasaki. (The second bomb, though more powerful than the first, fell off target; also, the city was surrounded by hills, so there were no firestorms.) Over the next five years another 140,000 people died, mainly from the continuing effects of radiation. No one knows exactly how many have died since 1950 as a result of the bombings, but cancer deaths attributable to radiation could number 100,000. Leukemia is thought to be on the rise among the estimated 370,000 survivors, and there's no way of knowing how many orphans or "orphaned elderly" have committed suicide or suffer lingering psychological damage due to loss of family or shock.

Death from radiation is not quick and it is not easy. After the bombs detonated in the Japanese cities, large numbers of victims suffered from nausea, vomiting, fever, anorexia, and diarrhea. A few weeks later their hair began to fall out; they ran high fevers; they noticed strange lesions, ulcers, and discolorations of their skin. Their noses and gums bled; there was also blood from internal wounds in their stools, urine, and sputum. The weakened victims died shortly of dehydration, blood loss, or infection.

The degree of genetic damage inflicted on the Japanese remains unclear because most mutations are recessive, or nearly so, and will not be evident for several generations. Exposure to many kinds of radiation causes many different kinds of chromosomal abnormalities, including mental retardation, stunted growth, shortened life-span, susceptibility to disease, and miscarriage and sterility. In animals the rate of mutations increases by 2 with each dose of 60 to 200 rads. Hence we can conclude that

---

nuclear war would probably introduce many mutations into the human gene pool, as well as temporarily or permanently sterilizing both men and women.

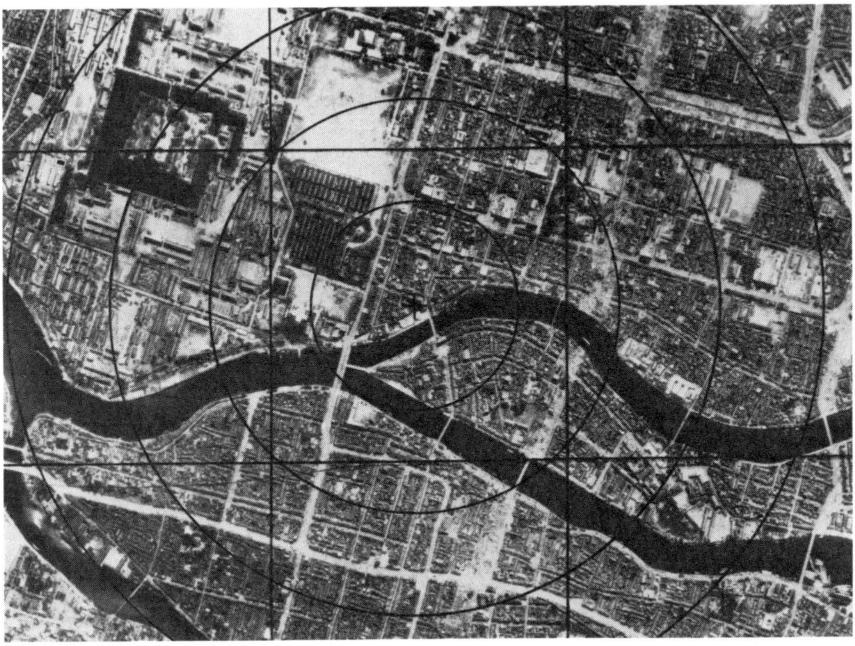

Ground Zero, as sighted from the air, before the atomic bombing of Hiroshima. The circles increase by radii of 1,000 feet. (Ground Zero is the spot directly below the explosion of the bomb.) Eighty-eight percent of the population within a 2-mile radius of Ground Zero died either instantly or within several weeks. One of today's 20-megaton bombs would flatten a city 8 miles from the blast center.

It is also clear that ecological chaos would result from a modern nuclear explosion, which would deplete the stratosphere's protective ozone layer. Blast would burn up nitrogen in the lower atmosphere, releasing enormous amounts of nitrogen oxides into the stratosphere and catalyzing the destruction of ozone. Ozone is necessary because it blocks ultraviolet light. Too much ultraviolet causes severe sunburn and blindness. The destruction of ozone could disrupt whole ecosystems. Insects that use ultraviolet light for guidance might be blinded or die, and plants that depend on such insects for pollination wouldn't survive. As vegetation died off, the soil would leach and erode.

THE BOMBING OF AN AMERICAN CITY

Ever since the bombings of Hiroshima and Nagasaki scientists have been trying to predict the consequences of nuclear attack on an American city. Studies in 1959 and 1962 considered the effects of a hypothetical attack on Boston, center of scientific, medical, and technological expertise at Harvard, MIT, and the electronics and computer firms along Route 128. Physicians for Social Responsibility (PSR), a group of doctors founded in 1962 to speak out against nuclear war, discussed the medical situation in postattack Boston and published their conclusions in a series of articles. In 1980, alarmed by the medical community's silence on the spread of nuclear technology and the buildup of arms, the group began to use the 1959 study as the basis for a series of symposia called "The Medical Consequences of Nuclear Weapons and Nuclear War" that doctors would present around the country. (Such symposia have since been conducted in Boston, New York, San Francisco, Seattle, Chicago, Albuquerque, Los Angeles, and other cities.)

PSR warns that the assumptions of the 22-year-old study of Boston are now quite conservative, because they involve the effects of a 10-megaton weapon. Though such a bomb is far more powerful than the Hiroshima bomb, which destroyed 3 square miles, one of today's 20-megaton bombs, if dropped on Detroit, would wipe out 70 square miles.

A Minuteman II missile would destroy 72 square miles, a Minuteman II, with Mark 12-A warheads, 88 square miles. The MX, with Mark 12-A warheads, would destroy 234 square miles, an area slightly larger than Chicago, and almost twice as large as Philadelphia.

If a 10-megaton bomb exploded at ground level, in downtown Boston, the spread of radiation would be maximized, and

This view of Hiroshima shows the total destruction resulting from the first atomic bomb. Today's bombs produce a thousand times the lethal blast, heat, and radiation, whose effects are shown here, in Hiroshima.

November 1, 1945. Interior view of atomic bomb damage to Hiroshima City Hall.

Blast buckled the columns of this wood frame building, which stood beyond the "fire fringe" at Hiroshima, Japan, 7,600 feet from Ground Zero.

October 7, 1945. Front entrance of the Hiroshima Gas Company building. Following the heat of an atomic explosion is a soundless blast wave that smashes everything standing.

November 4, 1945. Atomic bomb damage to the Hiroshima Prefecture Agricultural Association Building.

The terrific blast that followed the Hiroshima explosion on August 6, 1945, swept before it factories, office buildings, and churches. One year later, August 3, 1946, the euphoric caption on this U.S. Air Force photo reads: "Among the ruins new houses are being built and a new and better civilization is also being founded."

This steel frame building, The Shimomuna Watch Shop, had its first-story columns buckle under the terrific force of the blast, dropping the second story to the ground. Combustibles were destroyed by fire.

Bomb damage to the Hiroshima Telephone Company building resulted in complete destruction to the switch and relay room on the second floor.

Raging fire inside the Masasa Credit Association warehouse resulted in the spalling and disintegration of concrete walls.

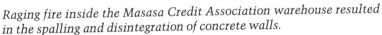

fire minimized. Nevertheless, aside from the massive bomb crater, half a mile wide and several hundred feet deep, everything within a radius of 4 miles, even heavily reinforced concrete buildings, would be wiped out, including most of the hospitals, clinics, and medical personnel in the Boston area. Within a 20-mile radius half the people would be killed or injured.

The detonation would release a tremendous amount of heat. More than 20 miles away survivors would suffer second-degree burns on all exposed skin, as well as burns from clothing, buildings, trees and shrubs on fire. Forty miles away people who looked at the fireball would be blinded by retinal burns.

The pressure wave from the blast, traveling faster than the speed of sound, would level or damage beyond repair all buildings up to 15 miles away. It would be followed by winds exceeding 1,000 miles an hour, fanning vast firestorms that would turn bomb shelters into ovens. Windblown fallout would kill people for hundreds of miles.

Many of the survivors would be badly burned, blinded, seriously wounded, or disoriented. As days and weeks passed, the problem of radiation sickness would spread. The sick and wounded would need immediate medical care, food, shelter, clothing, and water. But there would be no way to help them: hospitals, doctors, transportation, government, and society itself would have ceased to exist.

THE MEDICAL RESPONSE

The impossibility of meeting the survivors' needs, even in towns far from the blast center, which would be flooded with evacuees, make clear that no society can survive or cope with nuclear attack. Nowadays a single victim of severe burns or near-lethal radiation can occasionally be saved by the full resources of a highly specialized hospital unit, with countless transfusions and bone marrow transplants. But there is no way to cope with the hundreds of thousands of such cases, along with millions of people with lesser burns and injuries.

After a 20-megaton blast in Boston, the 1959 and 1962 studies estimated, the ratio of injured survivors to doctors would be over 1,700 to 1. In a prediction qualified as "unreasonably optimistic," the 1962 study said that

Even if a doctor worked 16 hours a day and spent only 15 minutes with each patient, it would still take up to 26 days for each casualty to be treated once.

In Boston there would be about 10,000 severe burn cases, unmanageable even if medical personnel and hospitals survived intact, since there are only 1,000 beds in the whole country for intensive care of burns. Large numbers of people would be suffering from lacerations and fractures, skin and retinal burns, and respiratory tract damage. Deaths from infection, starvation, dehydration, and acute and delayed radiation sickness would continue indefinitely. Fallout would have poisoned food and water. The surviving population would be reduced to roving hunters competing for food.

In every conceivable way a nuclear attack would be an unprecedented public health disaster. There would be more than 2.2 million dead bodies in metropolitan Boston, and they would have

"Well, I came to the street and I was very much surprised. There was no Hiroshima city in front of me. All the houses were flattened down. Hundreds and hundreds of people were walking in the street in their silence. Their shock was so great, they lost their words."—Tazu Shibama, a survivor.

to be disposed of to prevent epidemics. Many survivors would suffer permanent psychosis and neurosis, and there would be no one to treat them. Three months after the Hiroshima blast, many survivors were listless, lethargic, living in their own filth.

In March 1980, after the PSR symposium was conducted in Boston, attracting nationwide publicity, local citizens called the Boston civil defense office to ask what to do in the event of a nuclear attack. They were told to "go north." PSR believes that the government's civil defense program, based on the concept of evacuation, is as misguided and illusory as the 1950s' bomb shelter program. Today, while politicians emphasize that nuclear war can't be eliminated as impossible, civil defense planning remains virtually nonexistent.

After the Bostom symposium, Fred Haase, a health officer in the plans and preparedness division of the Federal Emergency Management Agency (FEMA incorporated Civil Defense several years ago), told *Science* magazine, "A nuclear attack would be a real catastrophe, no doubt about that . . . but if you take the worst situation, a lot of people are still going to survive." In short, civil defense has to plan for nuclear war. "The name of the game," Haase went on, "is the country that can plan now for survival and recovery is the one that's going to fare best in the long run."

But FEMA's civil defense planning, according to Haase, was in "shambles." Because FEMA is an umbrella organization that incorporated sections of HUD, DOD, and other agencies, as well as Civil Defense, it is riddled with fiefdoms and plagued by turf wars over budgets and priorities. In 1980 there was no public education going on, no training of medical professionals for emergency care of war victims, no medical-supply stockpiling, no evacuation planning except on paper. FEMA was agitating for a "national emergency health resources plan and policy"—to no avail. "We don't have a single up-to-date publication to mail out," Haase complained. And there was no government policy on evacuation, hospitalization, triage, or treatment. In 1973 the government scrapped a program providing equipment for setting up 2,500 emergency 200-bed hospitals.

A year and a half later Haase said the situation at FEMA was still "pretty murky." Planning policies continued to focus on evacuation. The agency was being reorganized, but as of the end of August 1981 there was no section devoted to health planning. "I'm the only remaining vestige of health activity," Haase said.

In November 1978 and November 1980, FEMA had participated in two joint exercises with the Pentagon, the first devoted to responses to "early phases of intense war," the second to "preconflict deployment and mobilization." The results are classified information, but Haase allowed that "in the last twenty years we've neglected attention to this area," and the exercises showed "our capability is not what we'd like to see it." (Since the accident at Three Mile Island, FEMA has considerably beefed up planning for civilian emergencies, but the section that covers evacuations from areas around nuclear power plants isn't involved in planning wartime evacuations.)

The lack of any real civil defense against nuclear attack makes it hard to argue with PSR's position that we are totally unequipped to deal medically with such an emergency. The Boston Study Group's data makes clear that no matter how many people might survive, the social fabric would be destroyed, and the long-term ecological, social, and genetic effects would be devastating.

Some members of PSR have gone on to form International Physicians for the Prevention of Nuclear War, which organizes meetings with senior physicians in other countries to focus world attention on the medical consequences of nuclear war in the hope of preventing it. So far, the group has been heartened by the fact that the Soviets seem to take the meetings very seriously. An IPPNW meeting in Washington in March 1981 drew medical leading lights from the United States, the Soviet Union, Great Britain, France, Japan, and elsewhere. The doctors see such international cooperation as the key to squelching the myth that civil defense would be of use in the wake of nuclear attack, that the survivors could be cared for adequately, that the worldwide effects would be negligible, and recovery swift.

> "At present, most of us do nothing. We look away. We remain calm. We are silent."
>
> Jonathan Schell
> *The Fate of the Earth*

Index

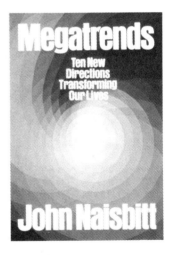